PRAISE FOR *PRO DESIGN AND THE CHAIN*

'Omera Khan has beautifully bridged the yawning gap between "product design" and "supply chain design" – a gap that has existed since the logistics/supply chain domain became an accepted management discipline over 50 years ago. Her emphasis on conscious "design" of the supply chain rather than the inside-out style of thinking that has prevailed to date is very refreshing. And she backs up her case with an impressive array of case studies. This book is long overdue, and a must read for leaders of all corporate functions.' **Dr John Gattorna, global supply chain thought leader and author, Adjunct Professor, University of Technology, Sydney, and Executive Chairman, Gattorna Alignment**

'In an accelerating world, this book provides an essential insight into the deep relationship between product design and the supply chain, and how many of the world's most successful businesses have used this intersection as a point of generating significant competitive advantage.' **Vikas Shah, MBE, Managing Director and CEO, Swiscot Group**

'A very thoughtful and insightful analysis identifying the complex requirements for designers to be aware of for their route to market and then identifying how to maximize their chances of success in their highly competitive marketplaces. The analysis is significantly enhanced through the use of highly relevant case studies.' **Michael Proffitt, former CEO of Dubai Logistics City**

'New products are the lifeblood of the business. However, the majority of new products never make it to market and those that do have a failure rate as high as 45 per cent. To be successful, new product development and commercialization must be cross-functional and cross-firm, and representatives from product design must be key members of the team. In this book, Omera Khan documents how product design can be used to improve the chances of launching successful new products while considering supply chain risks, the need for agility and the organization's sustainability goals. Numerous company examples are provided to illustrate how the integration of design and supply chain management were key to new product success.' **Doug Lambert, Mason Chair in Transportation and Logistics, and Director, The Global Supply Chain Forum, Fisher College of Business**

'An essential read for those wishing to understand the importance of design, its relationship with the supply chain and its impact on a business.' **James Dracup, Managing Director, Macnaughton Holdings**

Product Design and the Supply Chain

Competing through design

Omera Khan

KoganPage

First published in Great Britain and the United States in 2019 by Kogan Page Limited

2nd Floor, 45 Gee Street
London EC1V 3RS
United Kingdom
www.koganpage.com

c/o Martin P Hill Consulting
122 W 27th St, 10th Floor
New York, NY 10001
USA

4737/23 Ansari Road
Daryaganj
New Delhi 110002
India

ISBN   978 0 7494 7823 0
E-ISBN 978 0 7494 7824 7

British Library Cataloguing-in-Publication Data

A CIP record for this book is available from the British Library.

Library of Congress Cataloging-in-Publication Data

Names: Khan, Omera, 1976- author.
Title: Product design and the supply chain : competing through design / Omera Khan.
Description: London, New York, NY : Kogan Page Limited, [2018] | Includes bibliographical references and index.
Identifiers: LCCN 2018015884 (print) | LCCN 2018029586 (ebook) | ISBN 9780749478247 (ebook) | ISBN 9780749478230 (pbk.) | ISBN 9780749478247 (ebk.)
Subjects: LCSH: Business logistics. | Product design.
Classification: LCC HD38.5 (ebook) | LCC HD38.5 .K447 2018 (print) | DDC 658.5/752–dc23

Typeset by Integra Software Services, Pondicherry
Print production managed by Jellyfish
Printed and bound by CPI Group (UK) Ltd, Croydon, CR0 4YY

Shaan and Afra

CONTENTS

03   Product design and supply chain risk 63

04   Product design and the pursuit of agility 89

05   Product design and sustainability 123

LIST OF FIGURES

LIST OF TABLES

ABOUT THE AUTHOR

 Omera Khan has gained international recognition for her research and thought leadership in supply chain risk management. A passionate educator and highly acclaimed speaker, Omera is regularly invited as a keynote speaker at global conferences and corporate events. Inspiring and visionary, Omera is skilled at presenting complex subjects in a simple and thought-provoking manner. She provides training and workshops at all levels of the organization and specializes in strategic supply chain management, resilience and risk management, emerging trends and technologies, disruptive innovation and business transformation.

In addition to leading a number of ongoing research projects commissioned by government agencies, research councils and companies, Omera is an advisor to several universities developing courses in logistics, supply chain and operations management. She is currently a professor of Supply Chain Management and Director of the MSc in International Supply Chain Management at Royal Holloway, University of London (UK) and a professor of Supply Chain Management at the University of Agder in Norway and regularly teaches on an adjunct basis at universities around the world. Past appointments have included SP Jain (Singapore and Dubai Campus), Copenhagen Business School (Denmark), Manchester Business School (UK) and Massachusetts Institute of Technology (USA).

Omera's research has been published in leading academic and practitioner journals, several book chapters and she is lead author of the *Handbook for Supply Chain Risk Management: Case Studies, Effective Practices and Emerging Trends*. She founded and was Chair of the Supply Chain Risk and Resilience Research Club and the Product Design and Supply Chain Special Interest Group at the Logistics Institute, University of Hull.

Omera is a chartered fellow of the Chartered Institute of Logistics and Transport and the Institute of Operations Management and a steering committee member of the Institute of Logistics Research Network, a global network of academics and practitioners involved in state-of-the-art logistics and supply chain research. Omera was appointed as a board member of the executive committee of the CILT 'Leaders in Supply Chain' Forum, a group limited to the top 120 logistics and supply chain directors from Europe.

Omera began her academic career in the Centre for Logistics and Supply Chain Management at Cranfield School of Management. She joined the University of Manchester as a lecturer in Logistics and Supply Chain Management and then Hull University Business School where she held the positions of Senior Lecturer, Director of the Logistics Institute and Deputy Head of the Management Systems Group. Omera was awarded her PhD from Manchester Business School, an MPhil from the University of Manchester and a BSc in Textile Design from the University of Huddersfield where she specialized in woven fabric design and exhibited her designs at the Textile Centre for Excellence and the Business Design Centre. Omera was awarded first prize for the Cloth Workers Foundation award and commended for the National Wool Textile Export Corporation award.

Website: http://omerakhan.co.uk/

FOREWORD

For too long product design has been regarded as the preserve of specialists who are employed either for their creative or technical capabilities. This is the notion of design as an 'ivory tower' activity whose main purpose is to bring physical products to market whilst delivering both aesthetic and functional benefits to customers. Often this siloed approach has led to products being launched, which whilst being attractive to customers may also generate significant costs to the business throughout their lifetime thus potentially eroding profitability. Decisions taken at the drawing board stage will pre-determine ongoing costs in terms of manufacturing, distribution and product support. It is generally recognized that the majority of through-life costs of any product are already built in before it reaches the market.

It is 50 years since the philosophy of 'design thinking' first emerged as a concept with its emphasis on a more holistic approach to design, essentially viewing design as a problem-solving process which ideally extends across the organization. Whilst technical and creative specialists are still very much critical to the design process, in this broader approach they are augmented by inputs from across the supply chain – from procurement through to customer service and after-sales support.

In modern-day organizations a further impetus to the adoption of this end-to-end approach to design is the fact that complexity has increased in just about every aspect of their operations. Complexity arises because actions taken in one part of a system can have consequences – often unseen – elsewhere. One of the major drivers of complexity in a business context is product design. It follows therefore that by taking a more inclusive approach to product (and service) design the negative impact of complexity can be reduced. Design under this paradigm is no longer a stand-alone activity but rather a company-wide – or ideally supply chain-wide – concern.

This is the idea that underpins this original and ground-breaking book from Omera Khan. Whilst much has been written about product design and the innovation process generally, there has been little discussion about the relationship between design and the supply chain. This book helps to fill the gap. With the use of contemporary examples the author highlights the many ways in which customer value can be enhanced and complexity reduced by the adoption of an integrated approach to design and supply chain management. This is a book that deserves to be read by all those who have an interest not just in design per se but in how enterprises can succeed because the product design process and the supply chain are seamlessly connected.

Martin Christopher,
Emeritus Professor of Marketing
and Logistics, Cranfield University, UK

ACKNOWLEDGEMENTS

This book is the outcome of several years of research on a subject that has fascinated me since I first came to realize the significant role product design plays in the competitiveness and success of a firm – which years later I realized was through its supply chain.

As a child I visited international trade fairs with my father's company, procuring the next season's designs that would be developed into collections for home textiles. I was inspired and amazed by the negotiation with suppliers that produced our designs and in awe of my father who knew exactly what he wanted, knowing what would sell, and what wouldn't – or did he?

Years later after graduating with a degree in textile design I wrote my master's thesis on design procurement, followed by a PhD in supply chain risk management in the textile industry; my fascination on this subject only grew stronger.

Since then, I've had the good fortune to work with amazingly kind and inspirational people who have helped shape my research and formulate my ideas.

First of all, I owe enormous gratitude to Professor Martin Christopher, my mentor, who has guided and inspired me every day of my professional life. I have learned and developed massively through his encouraging spirit. His eye for detail, commitment and creativity helped to shape my thinking and develop my ideas with a confidence and wisdom I couldn't have imagined before meeting him.

To all my friends and family, who have given me strength when times were tough and the journey seemed bleak. My gratitude to Chris and Elsa Jephson for their endless 'book progress checks' and keeping me on track!

Thank you to all the companies that worked with me tirelessly and put up with all my questions and opinions. Particular thanks to Egil Moller Nielsen, for sharing his insights into Lego and ECCO,

Professor Chris Tang for contributing his case study on the Boeing 787 Dreamliner and to Daniel Sepulveda-Estay for sharing his experiences at Coca-Cola.

To all those wonderful people who have helped throughout the development and writing of this book, Malcolm Wheatley, Julia and the team at Kogan Page, thank you for your patience and time. The Design Council, and the CILT for the Seed Corn Fund in 2005 that helped elevate this research. I especially thank my students for their curiosity and imagination and for reminding me why I love teaching this subject.

Setting the design agenda

Introduction

Good design matters. Good designs – and good designers – can transform the prospects for a product, a business and even an industry. But while celebrating good design, and good designers, it is important not to lose sight of the supply chains that support those designs, and which transform them into saleable – and profitable – products.

This chapter highlights how good design makes a business more competitive and starts to explore how design issues go on to impact businesses in terms of their effect on supply chains' cost, **agility**, risk and **sustainability**. As we discover, design decisions can have implications in terms of issues such as time to market, **manufacturability**, **space utilization** and **transport intensity**, **inventory holdings** and **supply chain vulnerability**.

The gestation of the iPhone

Launched on 29 June 2007, Apple's iPhone was an immediate success. Combining a stylish form factor with an innovative glass touchscreen, it quickly became an object of desire, selling strongly.

Yet Apple wasn't a mobile phone manufacturer – or, in large part, a manufacturer at all, relying heavily on subcontract manufacturers such as China's Foxconn to turn its sleek designs into physical objects.

Instead, the iPhone had its roots in Apple co-founder Steve Jobs' dissatisfaction with a mobile phone design on which Apple had collaborated with mobile phone manufacturer Motorola, which along with brands such as Nokia and BlackBerry dominated the mobile phone market at the time. Intended to combine mobile phone functionality with Apple's innovative iPod MP3 player, the result was a product that the intensely design-conscious Jobs loathed.

'The ROKR [mobile phone],' wrote Jobs' authorized biographer Walter Isaacson, 'ended up having neither the enticing minimalism of an iPod nor the convenient slimness of [Motorola's original award-winning] RAZR. Ugly, difficult to load and with an arbitrary hundred-song limit, it had all the hallmarks of a product that had been negotiated by a committee, which was counter to the way that Jobs liked to work' (Isaacson, 2015).

Over a decade later, it is difficult to fully appreciate the revolution that the iPhone embodied. Right from the start, writes Isaacson, Jobs had vetoed the idea of having either a stylus or a keyboard – both deemed to be essential by experts if the new phone was to be a success. A plastic case was eschewed in favour of a slim case constructed of anodized aluminium, and for the screen a special high-strength scratch-resistant glass was commissioned from Corning Glass, involving an ion-exchange process that produced a compression layer on the surface of the glass.

Corning had developed the glass in the 1960s, but never found a market for it, and at the time, didn't have a factory capable of producing it. A series of conversations and meetings between Jobs and Corning Glass's chief executive Wendell Weeks persuaded Weeks to swing Corning behind the project, allocating its best scientists and engineers to the task.

The resulting iPhone wasn't cheap and at $500 was reckoned by competitors to be too expensive to be successful. Microsoft's then chief executive Steve Ballmer dismissed it, writes Isaacson. 'It's the most expensive mobile phone in the world – and it doesn't appeal to business customers, because it doesn't have a keyboard,' Ballmer told CNBC.

Instead, as Isaacson relates, by the end of 2010, Apple – which had never before produced a mobile phone – had sold 90 million iPhones and was reaping more than half the profits in the global mobile phone market.

By contrast, after several years of extensive losses, Motorola's mobile phone handset business was spun off into a separate division, Motorola Mobility, which was sold first to Google in 2011, and then (minus its extensive portfolio of technology patents) to Chinese electronics giant Lenovo. BlackBerry subscribers peaked at 85 million in 2013, falling to 23 million three years later, and after several years of heavy losses the company decided to launch Android-powered touchscreen-based smartphones (Tencer, 2017). After an equally troubled period, Nokia's mobile phone business was bought by Microsoft, in an eventually unsuccessful attempt by the software giant to boost its flagging Windows Phone business, which had just a low single digit share of the mobile phone market. At the time of writing, the mobile phone industry is dominated by Apple, and Google's Android ecosystem of phone manufacturers and software providers.

Design classics

Yet case studies such as the Apple iPhone aren't unusual. Time and again, we see innovative and stylish designs defy expectations to win commanding market positions.

Sir Alec Issigonis' Austin Mini, for example, which first rolled off the production line in 1959, quickly became a 1960s icon and became the best-selling British car in history with a production run of 5.3 million cars. Or the Volkswagen Beetle, which – although the prototype pre-dated the Second World War – was in mass production for 57 years, from 1946 to 2003. Or London's Routemaster double-decker bus, or the famous London black taxi.

In the home, similar examples of such 'design classics' include the Anglepoise 'balanced-arm' lamp designed in 1932 by British automotive designer George Carwardine; the multi-function Kenwood Chef food mixer, dating from 1950 and launched at the Ideal Home Exhibition in London that year; the horn-handled carving set; and the iconic 1950s Picquot Ware tea service, individually cast from a magnesium–aluminium alloy called Magnalium. The famous 1962 small-wheeled Moulton bicycle – a deliberate attempt to reinvent the bicycle in order to make it more user-friendly – came from Alex Moulton, an automotive engineer who had designed the suspension for the Austin Mini.

In the world of electronics and electrical appliances, too, there have been undoubted design classics. Sony's long-running Walkman series, for instance, combined portable stereo music with fashion, neatly capturing the zeitgeist of the 1980s. Meanwhile, the company's Triniton televisions coupled demanding levels of performance to stylish and desirable looks. So too with Dyson's famous 'bagless' Dual Cyclone vacuum cleaner, which became the fastest-selling vacuum cleaner ever made in the UK, outselling models sold by existing manufacturers, which inventor (now Sir) James Dyson had tried unsuccessfully to interest in his idea.

The world of fashion has many examples of design classics. Few would disagree with the widely produced cashmere twinset as a nomination, for example. Or the camel coat and trench, often associated with luxury fashion brands such as Jaeger or Aquascutum. DAKS tweed sports jackets are another example. So too is the wellington boot, often associated with Hunter Boot, which began life as the North British Rubber Company in 1856.

Design-led companies

Moreover, time and time again we see strong designs coming not as one-off hits, but from companies with a record of producing such designs. Design, you might say, is in their DNA.

To return briefly to Apple, for instance, Apple's iPhone came from a company design team with a long heritage of producing stylish 'must have' devices, stretching right back to the original Macintosh computer. Led by Apple's Chief Design Officer Sir Jonathan 'Jony' Ive, Apple has consistently turned out hit products such as the MacBook series, iPod, iPod Touch, iPad and of course Apple's flagship intuitive easy-to-use iOS operating system.

Iconic household appliance manufacturer Dyson, too, is a design-led company with – as at Apple, prior to Jobs' death in 2011 – a founder in a leading role as commanding visionary and serving as the company's public face. But although popularly styled as an inventor, Sir James Dyson is more properly a designer, with a background in industrial design and an education received at Byam Shaw School of Art and the Royal College of Art.

So too at Italy's Alessi, one of the world's leading manufacturers of designer kitchen and tableware, whose stylish products can be found on display at museums such as New York's Museum of Modern Art and Metropolitan Museum of Art, London's Victoria and Albert Museum and Paris's Pompidou Centre. Continuing a tradition begun under the leadership of Carlo Alessi, son of founder Giovanni Alessi, Carlo and his brother Ettore Alessi, Alessi puts design at the very heart of its business and has developed sophisticated processes for finding, commissioning and developing new designs from a world-wide network of talented designers and architects.

Family-owned Danish shoe manufacturer and retailer ECCO is another example of design excellence. Selling footwear and leather goods in almost 90 countries worldwide, vertically-integrated (an arrangement in which the supply chain of a company is owned by that company) ECCO controls every step of the production process from design and leather production to retailing, directly operating over 3,000 ECCO stores and 'shop in shops' (a designated space within a host retailer that is dedicated to a specific consumer brand, allowing them to sell goods under their own brand name), as well as selling through some 14,000 department, shoe and fashion stores. The company owns tanneries in the Netherlands, Thailand, Indonesia and China, and produces over three-quarters of its output in its own factories. Key to its success is its world-class R&D design centre in the Netherlands, where skilled leather craftsmen, designers and tech-nicians design not only finished products but also individual distinct leathers and tanning processes.

But it would be a mistake to conclude that design-centric compa-nies only ever design and manufacture physical products, and high-end luxury products at that. As we will see, while good design – and especially iconic designs – often command a price premium that takes them into the high-end or luxury category, the association need not imply a causal relationship. Consider the following design-centric companies, for instance:

- Denmark's LEGO, the world's sixth largest toy maker, has turned its expertise in manufacturing plastic blocks to exacting tolerances into becoming one of the world's leading brands and has transformed the processes of its design function in recent years. These changes have streamlined product development, and the processes developed by the in-house design function are now being used as a method to improve innovation across the entire business.

- Electronics, games and entertainment giant Sony has used design since the 1960s to differentiate its products and maximize the usefulness of its advanced technologies. The Sony Design Group across the world employs around 250 designers and has developed a set of core design values against which the company judges the success of all its products. Today, under Sony chief executive Kaz Hirai, Sony's heritage of simple clean designs and clear functionality are helping it to fight back against Apple and rising Korean manufacturer Samsung, whose products have taken market share from the iconic Japanese manufacturer (Walton, 2015).

- Virgin Atlantic Airways, founded in 1984 by British entrepreneur Sir Richard Branson, has innovation as a core brand value and uses design as a key competitive differentiator. The in-house design team manages many aspects of design for the airline, including service concepts as well as interiors, uniforms and airport lounge architecture, and works with a number of agencies worldwide.

- Microsoft, the world's leading supplier of operating system and office productivity software, has completed a significant evolution in its attitudes to design. Having once been a technologically driven organization, Microsoft now uses design thinking to focus on developing products that answer users' needs. With management support, this focus on user experience is also influencing Microsoft's organizational structure and culture.

- Whirlpool Corporation is a leading manufacturer of major home appliances. The Global Consumer Design unit at Whirlpool has a staff of over 150 people – industrial designers, usability experts, human factors experts, prototypers and materials specialists – and has developed expertise and processes that help the company respond to the demand for increasingly sophisticated and complex appliances, while at the same time developing products that are 'useful, usable and desirable'.

- Switzerland's Bodum Group is known for its stylish tableware and kitchenware ranges, including 'French press' coffee brewers (also colloquially known as 'cafetières'), vacuum coffee brewers, storage jars, water filters, cutlery and small appliances. 'Good design doesn't have to be expensive,' insisted founder Peter Bodum, and today that same philosophy guides Pia Bodum and Jørgen Bodum – respectively daughter and son of the founder, and the business's owners – as they manage the Bodum Design Group in Switzerland that develops the company's products.

In each case, such businesses have placed an importance on design that recognizes that good design should not be an afterthought but is central to the whole customer-facing proposition. They may go about it in different ways, but the goal is to consistently leverage design in order to add value to a product – or service – that goes beyond the merely functional.

What this book *isn't* about – and what it *is* about

This isn't a book about design. Or rather, it *is* a book about design, but not a book about *how* to design.

Reading it will not make you a designer if you are not one already. On the other hand, if you *are* already a designer, then it may well make you a better designer, or at least a designer who is more attuned to the commercial consequences of good – and bad – design.

Instead, this book is intended to address a readership of management professionals – especially, but not exclusively, manufacturing and supply chain professionals, and seeks to explain how good (and bad) design practices impact on corporate competitiveness.

Critically, this book shows how decisions taken at the design stage of a product's life cycle go on to affect that product's subsequent value to a company. Eighty per cent of a product's eventual supply chain costs are *already present* at the early stages of product design and development, meaning that design decisions can have significant through-life implications for risk, complexity and responsiveness. In other words, design is not just concerned with the appearance and functionality of products, it has an important role in determining the cost, pricing, risk and profitability profile of those products.

Illustrated by case study examples, this book therefore sets out to explain these linkages and show how design impacts:

- **Sales revenues,** by creating products for which demand is strong.
- **Pricing,** by designing products for which premium prices can be charged.
- **Time to market,** by designing products optimized for straightforward sourcing and scalability.
- **Manufacturing costs,** by designing products that can be easily and quickly constructed to a high standard.
- **Reliability,** by designing products that are rugged, durable and capable of being manufactured consistently.
- **Supply chain costs,** by designing products built from materials and components that lend themselves to straightforward sourcing from a competitive supply base, and cost-optimized logistics processes.
- **Supply chain risks,** by designing products built from materials and components that can be readily sourced from multiple (and preferably non-distant) sources.

- **Brand loyalty,** by designing products that exceed customers' expectations and provide them with a positive and compelling user experience.

- **Competitiveness,** by doing all of the above and creating a 'moat' that is difficult for competitors to sidestep or surmount.

Clearly, optimizing any one of these areas is a powerful force for good. But sadly, many companies don't actually optimize *any* of these areas. Some don't even try: the role of design, they believe, is to deliver a purely functional capability or deliver on purely aesthetic criteria. Some, though, do move beyond this, and design for one or two such areas – design for manufacturability, for instance, or design for supply chain resilience.

But it's a rare company that succeeds in tackling *most* or *all* of these areas, and truly harnesses its design skills to designing products that aim to enhance corporate performance right across the range.

Yet it can be done – and in this book you'll read how and learn what it is that businesses need to do in order to turn their design capabilities into a powerful force delivering compelling corporate competitiveness and performance.

And a taster, consider – once again – Apple. At the time of writing, according to the *Forbes* Global 2000, it is the number one company in the world in terms of market capitalization, with a value of US $586 billion; it is eighth in terms of sales revenues (US $233 billion); and number one (again) in terms of profitability (US $54 billion) (*Forbes*, 2016).

There are plenty of information technology companies out there producing hardware – the physical products that form the greater part of Apple's revenues. There are plenty of such companies with a reputation for innovation. But there are far, far fewer companies with a pedigree of investing in stylish design and of producing designs that carry forward the company strategy on such a broad footing. Among those, Apple stands supreme for its commitment to design as a competitive differentiator – and as its performance in the *Forbes* Global 2000 highlights, that commitment to design brings undoubted rewards.

What *is* design?

Let's now turn to (briefly) consider what we mean by design. For it is, unfortunately, a word with a number of sometimes conflicting meanings and connotations.

So too with the word 'designer', of course. Choosing a few scenarios at random, a designer might be someone conceiving a new wallpaper pattern, sketching out how your new kitchen might look, or bringing to life your new website or product catalogue.

In this book, we'll use the words 'design' and 'designer' simultaneously in a loose sense and yet also in a precise sense. Loosely, in that we are not going to preface these words with some kind of more precise qualifier – 'graphics designer', 'product designer', 'interior designer' or similar. And precisely, in that we mean design in the kind of painstakingly holistic sense that we have seen at Apple and in many of the companies briefly profiled in the previous section.

If we had to pick a single alternative to the words 'design' and 'designer', it would be '*industrial* design' and '*industrial* designer'. While not unknown in the UK and Europe, these terms are more commonly found in the United States, where industrial design is promoted and promulgated through the Industrial Designers Society of America.

As a membership organization, it has rather different goals to the UK's own Design Council, which was established as the Council of Industrial Design in 1944 to promote the value of industrial design in post-war Britain, but there are areas of overlap and common interest. One of those overlaps is a broadly shared vision of what good design constitutes, and why it matters.

'Design refers to much more than style, aesthetics or ergonomics: "design" is a way of thinking, not a way of making things look pretty,' is how the Design Council refers to design (Design Council, 2004).

A former chair of the Design Council, Sir George Cox, expanded on this in 2005 in a UK government-commissioned report – *Cox*

Review of Creativity in Business: building on the UK's strengths – which studied the links between business and creative professionals in the design, arts and related disciplines (Cox, 2005). 'Design is what links creativity and innovation,' wrote Cox. 'It shapes ideas to become practical and attractive propositions for users or customers. Design may be described as creativity deployed to a specific end.'

For its part, the Industrial Designers Society of America speaks of design in these terms: 'Industrial design is the professional service of creating products and systems that optimize function, value and appearance for the mutual benefit of user and manufacturer.'

A recent report from Warwick University Business School and the Design Council – *Leading Business by Design: Why and how business leaders invest in design* – further builds on this, making the following points about design, after studying design best practice in a number of leading businesses, and conducting interviews with business leaders from world-class companies such as Barclays, Diageo, Virgin Atlantic and Herman Miller (Design Council, Warwick Business School, 2014):

- **Design is customer-centred**: the benefit is greatest when design is intimately related to solving problems, especially customers' problems.

- **Design is most powerful when culturally embedded**: it works best when it has strong support in the organization, especially from senior management.

- **Design can add value to any organization**: design can benefit manufacturing and service-based organizations, small, medium or large.

For our part, and for the purpose of this book – bearing in mind that 80 per cent of a product's eventual supply chain costs are already present at the early stages of product design and development – three final points are worth making, especially when considering the impact of design on the supply chain.

- Design is a strategic tool that enables companies to differentiate products.

- Design is the creative process that impacts the extended supply chain from 'cradle to cradle'.[1]

- Design is an antecedent to identifying and managing risks early in the supply chain, where uncertainties are high but disruptions to the total supply chain can be kept to a minimum.

CASE STUDY   The pains of growth: Tesla's supply chain

Founded in 2003 by Silicon Valley tech-entrepreneurs and motor enthusiasts Martin Eberhard and Marc Tarpenning, Tesla found fame with its all-electric Tesla Roadster sports car, launched in 2008. Less than 10 years later – and producing just 76,000 vehicles a year, as opposed to the Ford Motor Company's almost 6.7 million vehicles – Tesla's stock market capitalization outstripped that of Ford, valuing the business at £38 billion (BBC News, 2017).

As with other disruptive entrants into existing marketplaces, a considerable part of Tesla's success can be attributed to willingness to rethink the art of the possible and disregard conventional wisdom. In Tesla's case, its approach to the development of its debut model, the Roadster, has been described as owing more to Silicon Valley software startups than Detroit.

Right from the start, the Roadster broke all the rules. The goal: zero to 60 miles per hour acceleration without a loss in torque; zero exhaust emissions; zero maintenance for the first 100,000 miles other than tyres; and a 250-mile range.

The cornerstone for this impressive performance was the power source for the electric motor, a lithium-ion (Li-ion) battery pack. The early Tesla engineering team designed a modular battery pack based on commercially available Li-ion batteries arranged in modules.

The battery pack consisted of an arrangement of 6,831 batteries, each slightly larger than a commercially available AA battery, embedded within an aluminium framework, and coupled to sensors for voltage, temperature and humidity, together with microprocessors and a cooling system.

Searching for financial capital to complete the production of a proto-type model, Eberhard and Tarpenning approached Elon Musk, another Silicon Valley entrepreneur, who had become a millionaire after the sale of PayPal, an internet payment company that he had founded. At the time, Musk was already engaged in SpaceX, intended to bring low-cost reusable rockets to the business of satellite launches.

With a working prototype and a receptive market, the Tesla team proceeded to build a supply chain in order to move to mass production. After being repeatedly rejected by several car manufacturers, Tesla Motors agreed in 2005 to purchase 2,500 vehicle 'gliders' (complete cars minus their powertrains) from Lotus Cars, a UK-based sports car manufacturer, at a maximum rate of 40 vehicles per week (Berdichevsky et al, 2006).

Although the supply chain that was built for the Tesla Roadster largely consisted of the already-established Lotus supply chain for the car chassis and accessories, the supply chain for the main technological feature of the car, its battery, was to be supplied by Tesla itself, as Lotus did not have previous experience with electric cars (Lotus formed its electric vehicles division only in 2008).

As there was no available battery pack in the market, Tesla had to identify the best way of manufacturing such a pack. At this time the Tesla engineering team had manufactured fewer than 10 battery packs and had built those by hand without using specialized tools or machinery. Given the labour-intensive production process for the battery pack, it was decided to outsource its assembly.

After a futile search for potential battery pack assemblers in China, Tesla selected a Taiwanese company with a production site in Thailand. Although this company had no previous experience in building batteries, it was skilled in aluminium forming for barbecue grills – which had a similar manufacturing process to the design of the battery packs – and could offer low labour costs.

The battery packs were considered dangerous cargo and thus were not allowed to be transported by air. Once assembled, the battery packs had to be shipped by sea freight to Lotus's manufacturing plant in the UK.

The assembled cars were then shipped to the port of New York, to be transported via land to California.

Although the contract with Lotus ran until December 2011, when production of the initial Roadster model was discontinued, it was soon clear that a better supply chain was required. Subsequent models of Tesla vehicles have been manufactured in an assembly plant in Fremont, California, formerly owned by a defunct joint venture between Toyota and General Motors. And a US $5 billion battery factory under construction in Nevada is expected to be the largest building in the world when fully developed.

Self-evidently, the process that Tesla followed for building its supply chain highlights the relevance of design for the supply chain in two particular ways:

- First, the design of the initial Tesla battery-pack supply chain was made with little consideration of the supply chain that would be required to scale up operations. This is a common feature of companies that undergo a rapid transition from concept development to prototype validation and then to large-scale production. The speed of scalability is generally highly dependent on the design, and can result in the difference between scaling up successfully and failure.

- Second, Tesla's initial supply chain configuration was derived from a Silicon Valley-style approach to manufacturing seen in companies such as Apple, Dell and Cisco: producing without actually having a manufacturing site, through the outsourcing of all manufacturing activities. However, in Tesla's case this led to an erratic supply chain configuration, dependent more on available offers for outsourced manufacture than on a serious attempt to optimize the selection and location of suppliers.

How design impacts competitiveness

By now, you'll have gathered that a central tenet of this book is that a company's design prowess and its competitiveness are inextricably linked.

Even so, the scale of that competitive outperformance might raise eyebrows. Consider a study from the Design Council in 2011, published to coincide with the government's innovation and research strategy for growth, which tracked 'design-intensive' publicly quoted firms over a 10-year period between 1994 and 2004 and compared them to poorer design-users. Through bull and bear markets alike[2] the design-intensive firms outperformed their peers by 200 per cent (Design Council, 2011).

More recently, the Design Council's in-depth *Value of Design* survey of 503 businesses estimated that every £100 that a 'design alert' business spends on design increases sales revenues by £225. The survey also reports businesses that see design as integral to what they do are more than twice as likely as others to experience rapid growth (Design Council, 2007). In a related finding , the Council reported in 2011 that 80 per cent of UK businesses believe that design will help them stay competitive in the current economic climate – a figure that rises to 97 per cent among rapidly growing companies (Design Council, 2011).

The *Cox Review* (Cox, 2005) argues compellingly that design impacts positively on practically every measure of business performance, including market share, growth, productivity, share price and competitiveness. In particular it showed that:

- Firms that ignore design have to compete mainly on price.
- 83% of design-led firms it looked at had introduced a new product or service in the previous three years compared to just 40% of UK companies.
- Of those design-led firms, 83% have seen their market share increase compared to the UK average of 46%.

- Design is integral to 39% of rapidly growing companies.

- 80% of the design-led firms that it examined had opened up new markets in the previous three years compared to the UK average of 42%.

- A business that increases its investment in design is more than twice as likely to see its turnover grow, compared to a business that does not do so.

Nor are the examples purely British. A report by John Bessant and Andy Neely of the Advanced Institute of Management Research – *Intelligent Design: How managing the design process effectively can boost corporate performance* – points to a number of corroborating formal academic studies (Bessant *et al*, 2006):

- A survey of 1,308 managers in Sweden revealed that companies with the greatest design maturity were found to enjoy very strong growth.

- In a study of 147 Dutch firms, integrating industrial design in new product development projects had a significant and positive influence on company performance.

- A study of 42 UK firms found statistically significant relationships between business success and various measures of long-term investment in design and innovation.

- A survey of 2,900 German service firms found that competitiveness was related to quality and flexibility in meeting different users' needs rather than on price.

But how exactly does this impact on competitiveness arise? What *exactly* is it that, say, the partnership between Steve Jobs and 'Jony' Ive delivered at Apple that has turned the company, as we've seen, into simultaneously the most valuable *and* the most profitable company in the *Forbes* Global 2000? Or, for that matter, delivered the competitive outperformance that design has had on any of the other companies that we have briefly profiled so far – companies such as Dyson, Virgin, ECCO and Bodum?

The clue lies in what those companies' design functions have been asked to deliver. Good-looking, attractive products, to be sure. But, as we've seen earlier, when thinking about the remit of this book, the impact of good design goes far beyond these.

Jennifer Whyte of Imperial College, London, together with a number of colleagues, argue that design may involve a number of aspects, all of which are covered by the term design (Whyte *et al*, 2005):

- Design for aesthetic appeal, such as styling, image or fashion status.
- Design for function, such as aircraft engines or Dyson's vacuum cleaners.
- Design for manufacturability, which involves thinking creatively about how the product can be effectively and efficiently manufactured.
- Design for sustainability, for example involving reuse and recycling.
- Design for reliability and quality in use.

Clearly, each of these will have an impact on overall competitiveness, as well as on factors such as pricing ability, brand loyalty, reputation and so on.

But a central tenet of this book is that the interface between design and supply chain management *also* has the potential to impact powerfully on competitiveness, should a business choose to take advantage of that interface.

In short, effective design – on its own – will, as we have seen, deliver powerfully on the competitive front. Good design results in a business bringing products to market that have customer appeal, provide some pricing power, deliver brand loyalty, reduce aftersales warranty and repair costs, and function effectively.

Moreover, effective design should also embrace design for manufacturability, in which designers give thought to how a product can cost-effectively be manufactured, further adding profitability and potentially delivering other benefits such as improved quality and reliability.

All of these things are good. But they don't, on their own, speak to a business's supply chain management capabilities or its **supply chain strategy**.

How might design influence those? Let's take a look.

Design and the supply chain

The term 'supply chain management' is relatively new. It was coined by Keith Oliver, a consultant at operations- and strategy-centric consultancy firm at Booz Allen Hamilton (now Strategy&) and was used by him in an interview with the *Financial Times* in 1982.

In the intervening years, it has come to be widely accepted and seen as an overarching term to describe and link a business's activities in areas such as manufacturing, purchasing, logistics and distribution, and forecasting and demand management.

Similarly, a business's supply chain for a given product can be thought of as the various suppliers supplying parts and raw materials going into that product's bill of materials – tier-1, tier-2, tier-3, tier-4 and so on – together with the various logistics processes that result in those parts and raw materials flowing forward to the end business in question. Then, once the product is manufactured, the customer-facing aspect of the supply chain sees the product reach the end customer, again moving through the appropriate logistics and distribution processes to do so.

Martin Christopher, an influential emeritus professor of supply chain management at the UK's Cranfield School of Management, defines supply chain management as the 'management of upstream and downstream relationships with suppliers and customers to deliver superior customer value at less cost to the supply chain as a whole' (Christopher, 2005). In a phrase that has since resonated widely, he has talked of competition as conceptually taking place between supply chains, rather than between individual businesses: just as the strengths of a business's supplier base are reflected in that business's own strength, so too will the weaknesses in that business's supplier base be reflected in its own weaknesses.

While the phrase 'supply chain management' is now widely used, some argue that a more accurate phrase could be 'demand chain management' in order to reflect the fact that the chain should be driven by the market, not by suppliers. Equally, there is a view that the word 'chain' should be replaced by 'network' since multiple suppliers and customers are involved in the extended enterprise. Moreover, this network-centric approach is reinforced by the fact that each organization in a supply chain is to some extent dependent on other organizations within the supply chain or network, with a resulting need for companies to integrate activities. In a still more recent development, the term 'value chain' is now being used as well.

Irrespective of the nomenclature, though, it is clear that design can have an influence on the efficient operation of the supply chain, and that design decisions can have significant through-life implications for risk, complexity and responsiveness. Indeed, as we will demonstrate in this book, in a very real sense a business's supply chain begins on the drawing board, with design decisions potentially impacting the cost, risk and profitability profile of the business dramatically.

Consider, for instance, the following ways in which design decisions can impact the supply chain:

- **Time-to-market and time-to-volume:** decisions on the functionality of products can increase manufacturing complexity and reduce flexibility and responsiveness.

- **Manufacturability:** similarly, product designs can influence manufacturability, impacting manufacturing costs, reject levels, lead times and quality levels.

- **Space utilization and transport intensity:** the physical characteristics of the as-designed product and its packaging will affect the utilization of space in storage and transport.

- **Added complexity through lack of commonality**: decisions on product design impact the bill of materials, with low levels of commonality adding complexity, adversely impacting inventory holdings and reducing the potential for economies of scale to reduce as-purchased costs.

- **Increased replenishment lead times**: some design decisions can influence the choice of supplier and therefore increase lead time.

- **Supply chain vulnerability**: sourcing products from offshore suppliers can potentially increase supply chain disruption.

- **Sustainability and carbon footprint**: 'design-for-sustainability' and end-of-life considerations have now become an important priority for most businesses.

- **Late stage customization**: the ability to postpone the final configuration or packaging of a product – thereby influencing customer lead time and inventory holdings – will clearly be enhanced or constrained by design decisions.

- **After-sales support**: for those products requiring after-sales support through service parts, the design of the product can clearly have implications for inventory levels.

Clearly, these are all important issues, with an obvious impact on business performance metrics at the highest level. And given – as a rough rule of thumb – that 80 per cent of a product's supply chain costs are determined at the design stage, the obvious question is easily stated: what can business do about it?

CASE STUDY   ECCO: balancing cost and complexity

Founded in 1963 by Karl Toosbuy, Danish shoe manufacturer and retailer ECCO sells footwear and leather goods in almost 90 countries worldwide, directly operating over 3,000 ECCO stores and 'shop in shops', as well as selling through some 14,000 department, shoe and fashion stores.

Noted both for its design excellence and its commitment to a vertically integrated business model, ECCO owns leather tanneries in the Netherlands, Thailand, Indonesia and China, and produces over three-quarters of its output in its own factories. Given the labour-intensive nature of shoe manufacture, these are located in low-cost economies such as Slovakia, Portugal, Thailand, Indonesia and China.

At its R&D and design centre in the Netherlands, skilled leather craftsmen, designers and technicians design not only finished products but also create individual distinct leathers and tanning processes. Described internally as 'From Cow to Shoe', it is a model that is designed to provide ECCO with full control over the quality of both its products and the brand experience. Strategically, too, leather is the most expensive component of a leather shoe and operating its own leather tanneries – processing raw hides into various finished leather types – also provides ECCO with an advantage over competitors who have to buy leather from third-party suppliers.

Most obviously, one benefit of this vertical integration is that ECCO's R&D and design teams can work closely together with the firm's tanneries in developing new colours and leather types. But the vertical integration also serves to reinforce ECCO's reputation and brand image in respect of quality: as ECCO does not operate in the fast-fashion segment of the market – indeed, some 'classic' shoe models are decades old, and most are in production for at least one season every year– a long vertically integrated supply chain is an affordable luxury, as quality is seen as being more important than being up to date with the latest fashion.

ECCO has two seasons, a Spring/Summer season and an Autumn/Winter season, each of which in manufacturing and logistics terms is characterized by a 'main order', planned some four to six months ahead in order to optimize manufacturing costs, and a series of subsequent replenishment orders to reflect higher than anticipated sales of particular shoe models, sizes or colours. Master production planning in the manufacturing sites plans these main orders into production batches for each shoe design, broken down by model, size and colour, based on the main orders on hand, and a supplemental level of production that includes an element of seasonal demand and anticipated replenishment orders.

Prior to 2002, retailers placed their indicative main orders several months ahead for the models and colours that they wished to sell, and the assortment of shoes that was then shipped to them reflected a centrally determined typical shoe size profile for a given market. In Scandinavia, for instance, size 43 is the most commonly required shoe, so there would be more of that size in a given assortment than any other. Each main order assortment would then be broken down into phased deliveries, spread over the season.

Again, prior to 2002, these assortments would be packed into cases at ECCO's production facilities and then shipped to the company's distribution centres for storage until delivery took place. However, in the event that retailers' actual updated orders did not correspond to retailers' original orders – placed four to six months earlier, remember – ECCO's distribution centres then had to break open two case-packs, and repack the two cases' sizes and colours into a new case-pack that matched the actual order, made up of main-order shoes and replenishment orders. The remaining sizes and colours from the two cases would obviously not match any predefined assortment, and were therefore restocked as single SKUs.

This repacking added value to neither the product nor the retailers and was consuming valuable distribution centre capacity: service levels to stores were in the region of 70 per cent, and an internal study in 2001 showed that over 50 per cent of all case-packs destined for retailers were being repacked. Clearly, if repacking could be reduced or eliminated it would bring significant value to ECCO in terms of reducing costs, stripping out unnecessary activity and improving service levels to stores.

The solution, implemented over 2002–2004, was to split the supply chain into two separate streams – one for main orders and the other for replenishment orders. In order to execute this new concept, facilities known as production distribution centres were constructed adjacent to the company's manufacturing units, with manufacturing no longer packing shoes in assortment cases, but moving them in bulk to the production distribution centres next door to each plant.

As main orders were readied for delivery, the production distribution centres picked and packed shoes to the latest main-order information, generally some 40–50 days prior to the requested delivery date. This meant that there was no repacking taking place, as picking and packing was done from bulk storage. By definition, each shipment reflected the most up-to-date required size and colour order profile known to ECCO.

The 40–50 days – derived by back-scheduling from the requested delivery time by the length of the transport lead-time from the production distribution centres – was obviously longer than was previously the case, as the production distribution centres were located at ECCO's manu-facturing sites in Slovakia, Portugal, Thailand, Indonesia and China, and ECCO's main markets were in North America and Europe. Here, 'break bulk' cross-dock 'hubs' received the shipments from the manufacturing sites and despatched them to the individual retailers for which they were destined.

Stock for replenishment orders, meanwhile, was shipped in bulk to a series of *replenishment* distribution centres, which in this case were located on the same continent as the retailers to which the shoes in question were destined to be shipped, ie, Europe, North America, Australia, Asia and so on. Replenishment orders placed by retailers could then be picked, packed and shipped within 24 hours.

The changes to the supply chain had a significant impact on the way that master production planning was performed and also on the processes of sales forecasting, inventory planning, order intake and fulfil-ment. Moreover, the project impacted everybody in the organization, all the way from manufacturing through to retailing.

But the effort was clearly worthwhile, even though the new supply chain was more complex to manage than the previous model. Customer service levels, for instance, went from a level of around 70 per cent to over 95 per cent, while logistics costs declined significantly from a level of over 14 per cent of sales, to less than 9 per cent. Fifteen years after it was first developed, ECCO's new supply chain model was still in use.

Developing a shared agenda: design and the supply chain working together

In essence, this book argues that design decisions, as related to the supply chain, impact the business in four key areas:

- **Cost,** via manufacturing costs and supply chain costs. From this it also follows that there will be an impact on profitability.

- **Agility,** via impacts on product and inventory complexity, choice of suppliers and lead times.

- **Risk,** via complexity, sourcing decisions and lead times.

- **Sustainability,** via the product's (and the business's) resource footprint, its impact on remanufacture and recyclability of products, and through space utilization and transport intensity.

Given this, the argument follows that the supply chain 'begins on the drawing board', and it is important that the entire organization sees that design decisions affect the business in respect of these four areas. So it is crucial that design is perceived as much more than just an activity that creates novel ideas or brings stylistic changes to products. There is in fact a more strategic role for design, which impacts the total supply chain.

Traditionally, this conceptual – and organizational – linkage has not been generally present. While some organizations have seen the link and exploited it to the full, many have done so either patchily or not at all.

Nor is it difficult to see why this has occurred. Historically, most organizations have been functional in their structure, with responsibility for each stage in the value chain – including design – being separate from all other stages.

One inevitable result of this is that influence over design, and the design function, has been fragmented. Procurement, manufacturing, distribution, risk management – no single voice has been strong enough to adequately influence the design process. The emergence of a single unified supply chain management function not only brings a muscular new voice to the debate, but also acts as a prioritization mechanism: rather than dealing with competing issues and concerns put forward by procurement, distribution, manufacturing and so on, it is possible to create a single, ordered, 'shopping list'. Here, do *this*.

Another issue is that in many businesses the new product development process has not only been organizationally separate from the rest of the business, but also linear and sequential, with a consequent impact on time to market, and a tendency to consider the manufacturing and supply chain aspects of a product's design as an afterthought, if at all. Put another way, the product has been conceived and developed first, and optimized subsequently.

Partly, that has been informal: 'here's the product – what do you think?' But partly it has been formal: in some organizations, it is the role of manufacturing engineers to take finished product designs and then re-engineer or redesign them for manufacturability. Likewise, packaging engineers do the same for packaging and procurement engineers do it for procurement. These 'after the event' approaches, while better than nothing, are inherently wasteful and suboptimal: why not simply get it right first time, therefore speeding – not slowing – time to market?

This book contests that in today's challenging markets, these 'silo' type organization structures and 'after the event' fixes and kludges are no longer fit for purpose, especially when coupled to development processes that are themselves linear and sequential. Things have to change.

And the nature of that change is clear: in contrast to today's separate 'silos', the design function and the supply function need to develop a shared agenda – one that addresses these various issues of cost, agility, risk and sustainability.

In this book, we look at how to shape that shared agenda and how to build opportunities for shared discussions to take place:

- in Chapter 2, we look at the interface between product design and the supply chain;

- in Chapter 3, we look at design risk management;
- in Chapter 4, we examine design and agility;
- in Chapter 5, we look at design and sustainability;
- and in Chapter 6 we look at how the themes in each chapter link together, how closer links between the supply chain and design functions can be achieved in practice, and discuss the broader implications for education and development.

Checklist

Questions for design professionals

☐ Do you know your business's five most significant supply chain challenges?

☐ To what extent might product design be capable of influencing these challenges?

☐ To what extent do your business's design briefs formally stipulate manufacturing and supply chain considerations as design outcomes?

☐ How frequently (if at all) do you meet with your colleagues in the supply chain function?

☐ Are there formalized joint meetings with the manufacturing and supply chain functions?

Checklist

Questions for supply chain professionals

☐ Do you know where your business's design function is located?

☐ How far away is it from the location of your business's principal manufacturing and supply chain activities?

☐ Are there formal mechanisms for manufacturing and supply chain professionals to provide input to the design function at the early stages of a product's design?

☐ Do you ever have formal meetings with the design function to discuss manufacturing and supply chain issues?

☐ Can you think of products within your business where seemingly simple design changes could make a significant impact on supply chain outcomes?

Notes

1 According to William McDonough and Michael Braungart, Cradle to Cradle is a design framework for going beyond sustainability and designing for abundance in a Circular Economy.

2 If a person is optimistic and believes that stocks will go up, he or she is called a **bull** and is said to have a 'bullish outlook'. A **bear** market is when the economy is bad, recession is looming and stock prices are falling.

References

BBC News (2017) Tesla's market value overtakes Ford [Online] available at: http://www.bbc.co.uk/news/business-39485200 [accessed 11 January 2018]

Berdichevsky, G, Kelty, K, Straubel, JB and Toomre, E (2006) The Tesla Roadster Battery System [Online] available at: http://large.stanford.edu/publications/power/references/docs/tesla.pdf [accessed 11 January 2018]

Bessant, J, Neely, AD, Tether, B, Whyte, J and Yaghi, B (2006) *Intelligent Design: How managing the design process effectively can boost corporate performance*, Advanced Institute of Management Research, London

Christopher, M (2005) *Logistics and Supply Chain Management: Creating value adding networks*, 3rd edn, Prentice Hall, London

Cox, G (2005) *Cox Review of Creativity in Business: building on the UK's strengths*, HMSO [Online]available at: http://webarchive.nationalarchives.gov.uk/20120704143146/http://www.hm-treasury.gov.uk/d/Cox_review-foreword-definition-terms-exec-summary.pdf [accessed 31 January 2018]

Design Council (2004) *Design in Britain*, Internal report by the Design Council

Design Council (2007) *The value of design factfinder report* [online] available at: https://www.designcouncil.org.uk/sites/default/files/asset/document/TheValueOfDesignFactfinder_Design_Council.pdf [accessed 15 May 2018]

Design Council (2011) *Design for Innovation* [Online] available at: https://www.designcouncil.org.uk/sites/default/files/asset/document/design-for-innovation.pdf [accessed 11 January 2018]

Design Council, Warwick Business School (2014) *Leading Business by Design: Why and how business leaders invest in design* [Online] available at: https://www.designcouncil.org.uk/sites/default/files/asset/document/dc_lbbd_report_08.11.13_FA_LORES.pdf [accessed 11 January 2018]

Forbes (2016) *Forbes* Global 2000 [Online] Available at: https://www.forbes.com/global2000/#7e1deab2335d [accessed 11 January 2018]

Isaacson, W (2015) *Steve Jobs*, Abacus, London

Tencer, D (2017) BlackBerry's Market Share Declines to 0.0%, Huffington Post Canada [Online] available at: http://www.huffingtonpost.ca/2017/02/16/blackberry-market-share_n_14798544.html [accessed 11 January 2018]

Walton, M (2015) Inside Sony Design: How a new breed of designers hope to fix the company's fortunes, Ars Technica [Online] available at: http://arstechnica.co.uk/gadgets/2015/10/inside-sony-design-how-a-new-breed-of-designers-hope-to-fix-the-companys-fortunes/ [accessed 11 January 2018]

Whyte, J, Bessant, J and Neely, AD (2005) Management of Creativity and Design within the Firm, paper commissioned by the Department of Trade and Industry as an input to the Creativity Review [Online] available at: http://webarchive.nationalarchives.gov.uk/20111108232833/http://www.bis.gov.uk/files/file14795.pdf [accessed 11 January 2018]

Further reading

Baer, D (2014) The Making of Tesla: Invention, betrayal, and the birth of the roadster, *Business Insider*, [Online] available at: http://uk.businessinsider.com/tesla-the-origin-story-2014-10 [accessed 11 January 2018]

Fine, C, Padurean, L and Werner, M (2014) The Tesla Roadster (A): accelerated supply chain learning, Massachusetts Institute of Technology Sloan School of Management [Online] available at: http://asb.edu.my/wp-content/themes/asb/Tesla%20Roadster%20(A)%202014.pdf [accessed 11 January 2018]

The interface between product design and the supply chain

Introduction

The supply chain begins on the drawing board: as we will see, that is the central contention of this book. This chapter explores the links between the design function and the supply chain function and shows how design decisions can have a significant effect on supply chain – and therefore business – outcomes.

Looking in detail at fashion retailer Marks & Spencer, computer manufacturer Dell, aircraft manufacturer Airbus and 'fast fashion' retailer Zara, it contrasts **constrained design** with unconstrained design and introduces the '4C' best-practice design model. Finally, it briefly outlines the application of product life cycle management technology as one means of facilitating better cooperation between the design and supply chain functions.

The decline of a high street icon

For years, it had seemed as though iconic British fashion, homewares and food retailer Marks & Spencer could do no wrong. In 1998, it became the first British retailer to announce annual pre-tax profits of over £1 billion and under long-serving chair Richard

Greenbury – a former Marks & Spencer management trainee who had joined the business at 17 – it seemed set to continue dominating the high street.

But change was afoot, and that same year would later be seen as Marks & Spencer's high-water mark. The company had adapted badly to trends such as globalization, casual clothing and accelerating fashion life cycles. Over three-quarters of its products were sourced from the UK and despatched to retail outlets from warehouses holding an inventory equivalent to seven weeks' sales.

A complicated value chain saw product ranges planned and set in train by a series of headquarters-based design and purchasing teams a whole year before delivery to store, with replenishments often being slow to arrive due to poor forecasting and planning.

Even as the company celebrated breaking the billion pound profit barrier, observers were pointing out that Marks & Spencer seemed increasingly out of touch with both the times and the marketplace. So it was to prove. Michael Pich, Ludo Van der Hayden and Nicholas Harle write the following about Marks & Spencer's fashion predictions (Pich *et al*, 2002):

> Convinced that grey and black would be in fashion during the 1998/1999 season, M&S had developed its entire collection around those two colours. Due to the lead times in its value chain, [the company] had to make this decision one year in advance of the season. Regrettably, it lost its bet. 'When we realised that our choice was wrong, it was too late to order more colours,' commented an M&S spokesperson.

The result was what was described in the business press as 'the biggest off-season sale in [the company's] history' – with such a sale being itself a rare event at Marks & Spencer. Prices were slashed on over £500 million of goods, pre-tax profits almost halved to £656 million and a precipitous decline in the share price began (BBC News, 1999).

Forced by circumstances to break with tradition and actually advertise its wares, the company also found itself departing from its long-held practice of refusing to accept third-party credit cards, with its share of the UK clothing market falling to a then all-time low of 14.3 per cent. Meanwhile, a survey showed that its customer base was ageing: in its core UK heartland, the company had only a 5 per cent market share among 15–24-year-olds, but a 24 per cent market share among those 65 and older.

A management shake-up saw 31 of the company's top 125 executives leaving the firm, Greenbury becoming chair and newly-appointed chief executive Peter Salsbury vowing to turn the business around in two years.

'The company is now going to start buying much closer to the seasons, as well as improving its styling and quality, and sharpening up its prices,' observed respected retail analyst Richard Hyman. But the ensuing supply chain restructuring, concluded news magazine *The Economist*, had been 'botched'. 'It upset suppliers. Deliveries coming from farther and farther away became harder to control. Distribution costs rose. Quality fell' (*The Economist*, 2001).

Greenbury left the business in 2000, prompting the appointment of Belgian consumer goods and retail executive Luc Vandevelde as chair. Salsbury departed not long afterwards and in an interview with the *Daily Telegraph*, Vandevelde conceded in 2001 that the company no longer 'understood' its market and had lost its core clients in the 35 to 55 age range (Jackson-Proes, 2001).

Almost incredibly, history then repeated itself. After years spent ignoring the impact of globalization, the company had finally joined its competitors in overseas sourcing in the late 1990s. But Christmas 2003 went badly wrong for it. As usual, it had placed its orders for winter fashionwear many months previously – although, in most cases, earlier than the full year ahead that had previously been the norm. But the company's bet that the coming winter would be cold and dry, leading to strong demand for knitwear, misfired: instead, the winter of 2003 was warm and wet.

Worse, back in the spring, the company had – naturally – been unable to predict England's heady success in the Rugby World Cup, which kept shoppers of both sexes glued to television sets and away from its stores. Consumer footfall in the critical run-up to Christmas was low – much lower than normal – and those shoppers that did step through the doors were then finding the 'wrong' ranges on the shelves. Bloodied, the company launched its first ever pre-Christmas sale, deeply discounting prices in an effort to clear stocks.

The design-centric business

Why the focus on Marks & Spencer and these now long-ago events? Because, quite simply, Marks & Spencer's turn of the century travails epitomize everything that this book has to say about the linkage between product design and the supply chain in design-centric businesses such as fashion retailers.

At its peak – and those 1997–1998 results still mark an unsurpassed high point in the company's financial performance, even today – Marks & Spencer was reckoned to be the world's second most profitable retailer (after Walmart), with an enviable reputation for combining commercial prowess with shareholder returns and corporate and social responsibility.

But the world changed. No longer able to impose a Henry Ford-style view of fashion on its customers ('You can have any colour you like, as long as it's black'), it found itself outsmarted by competitors who didn't place orders a year in advance, gambling that their choices would be correct, but instead responded nimbly to what was actually selling in the marketplace.

Put another way, as we saw in Chapter 1, the supply chain begins on the drawing board, with the design function and the supply chain function working together to pursue a shared agenda – an agenda aimed at maximizing responsiveness and choice, while minimizing cost, inventory and waste. Waste? Very much so: consider the waste in terms of revenue and profit represented by 'fire sales' of unsold goods that customers must be incentivized to purchase.

And put yet another way still, by bringing the product design function and the supply chain function closer together, design-centric businesses have an opportunity to *minimize product cost* and *supply chain cost*, and *maximize supply chain agility* while also *minimizing supply chain risk*.

Design-centric: not just for fashion

Nor is this an issue solely for fashion firms – although with its short product cycles and ever-changing consumer tastes, fashion is an industry that almost uniquely serves to highlight both good and bad practice at its starkest. Repeatedly in this book, we will see how fashion-related businesses show the way in linking product design to their supply chains – and not only that, but doing so in a very conscious and deliberate way, often placing that link at the centre of their business strategy.

As an extreme example of that linkage at work – and especially the business strategy dimension – in a design-centric business other than fashion, consider the emergence of Dell Computer Corporation as a disruptive force in the personal computer industry.

Founder Michael Dell famously dropped out of college to found the company in 1983, when he saw that he could buy up surplus computer parts, add them to bulk purchased basic IBM machines and then sell the souped-up machines at a price that had people clamouring to buy, especially when coupled to an ability for purchasers to specify exactly which additional components were added to the basic machines.

The transition from souped-up versions of other people's machines to his own line of computers occurred in 1985, again predicated on a business model of selling directly to customers, rather than going through dealers, and again based on offering customers the ability to specify exactly what features and components they wanted their built-to-order computer to contain.

In the marketplace, the ability to customize a computer to a customer's own requirements in terms of particular configurations of memory, disk drives, processor and storage technologies was a strong selling point. But for Dell, designing a product, business model and supply chain that was oriented around a build-to-order business model delivered enviable (and market beating) performance – especially once Dell began selling its machines online in 1996 (Wheatley, 1998).

Place an order online for a Dell computer and a remarkable sequence of events would spring into action. Almost immediately, software systems worked out what parts were required to build the personal computer in question. Then, the inventory systems of suppliers were interrogated, in order to make sure that the parts were in stock before placing an order for them. And within a couple of hours, the parts would have been delivered from suppliers' nearby just-in -time 'hubs', enabling assembly to commence.

It was an impressive feat, compounded by a financial performance that was even more impressive: minimal levels of work-in-progress, no raw materials or components inventory at all and a finished goods inventory that consisted chiefly of spare parts for service. Moreover, the Dell approach was well suited to the personal computer industry: as the price of memory chips and processors fell, the company's exposure was limited to little more than the computers on its assembly lines. Best of all, the system served to ensure that the company's rapid growth was largely self-financing.

In 1999, Dell surpassed Compaq to become the largest personal computer manufacturer in the world and was within two years of pulling off the same feat in the manufacture of Intel-based servers.

'We offered mass customisation versus the "one size fits all" approach of the market leaders: customers liked that,' observed founder Michael Dell (Wheatley, 1999), contrasting Dell's own lean and flexible production lines with the highly automated and rigidly standardized lines of his competitors.

Put another way, Dell's genius was to see the potential to not only close-couple the links between product design and the supply chain, but arguably blur the barrier between them to the point of non-existence. By making its machines easily configurable as they travelled down the assembly lines, Dell transformed much of the design process into a configure-to-order process carried out by the end customer, with menus of options cleverly linked to a just-in-time replenishment process operated by suppliers. Equally cleverly, those suppliers, as well as the customer-selectable components themselves, were selected with a weather eye for just this business model.

To some, the Dell example will be seen as extreme. But it is in fact simply an extension of the entire pull-based, demand-driven supply chain logic that businesses have been striving for since just-in-time and lean production methodologies arrived in the West in the 1980s. Why hold warehouses full of computers that customers might not want, when it is possible to design computers capable of being customer-configured to whatever specifications customers *do* want, backed up by a business model fine-tuned to deliver those very computers?

Airbus's connection failures

If Dell Computer is an example of best practice in a non-fashion context (albeit one not necessarily applicable in every situation), let's now turn to look at an example of well-intentioned practice going wrong.

Specifically, let's look at the development of Airbus's giant Airbus A380, the 'double-decker' passenger aircraft that is not only the largest passenger aircraft in the world, but which also represented a huge step change in design and production complexity compared to Airbus's earlier designs. Moreover, the design and development time-table from project sign-off to first aircraft delivery to customers was just six years – a timetable seen by Airbus as aggressive but doable.

Yet with major structural sections of the A380 being built in France, Germany, Spain and the United Kingdom, and then finally assembled into a finished aircraft in Toulouse, France, it was obvious that close cooperation and coordination between designers, and between designers and production engineers, in each of Airbus's 16 facilities involved in the Airbus A380 was going to be even more important than usual.

What's more, development and production of the Airbus A380 was to proceed concurrently, using a proprietary Airbus method-ology, Airbus Concurrent Engineering, yet with design staff in the various facilities taking different approaches and – critically – using different design software.

In particular, while French designers used two three-dimensional computer modelling programs called Catia and Circe, that had been developed by the French software firm Dassault Systèmes (and deployed with great success on Boeing's 777 airliner a decade previously), German engineers preferred to work with a two-dimensional package developed by US software firm Computervision, that had been popular in the 1980s.

The result: sections of the aircraft's rear and front fuselage arrived at Toulouse to be assembled into complete fuselages and the cables installed in them were too short: quite literally, the aircraft couldn't be put together. And with 100,000 different wires totalling 330 miles, in each aircraft, the size of the headache was obvious. As increasingly incomplete fuselage sections arrived at Toulouse, an army of German assembly mechanics arrived on site, outnumbering the French workforce (Clark, 2006).

Three separate delivery delays were announced, totalling two years. The scheduled ramp-up of full production volumes slowed, with just 13 aircraft projected for delivery in 2008 and first delivery taking place in October 2007. The projected earnings shortfall through to 2010 climbed to €4.8 billion and a series of high-level management departures followed.

No magic wand

Clearly, there are various themes emerging from these case studies of design-centric businesses such as Marks & Spencer, Dell and Airbus. Consider, for instance, some of the most important ones:

- Most obviously, the linkage between product design and the supply chain *matters*. The linkage isn't an optional, nice-to-have sort of thing: instead, it is an important determinant of business success. If a business designs the products that it sells, then – as we have seen in Chapter 1 – product outcomes in terms of cost, risk, time-to-market, agility and sustainability are all impacted by the calibre and nature of that linkage.

- Moreover, the downsides of getting this linkage wrong can be career-limiting, as at Marks & Spencer and Airbus. Business rewards success and penalizes failure, and a close link between design and the supply chain helps to orient businesses towards the former, not the latter. Those affected by failure (and rewarded by success) aren't simply the high-profile executives at the top, but the entire organization, both within the design and supply chain functions and outside them.

- Achieving this link between product design and the supply chain function isn't necessarily a simple matter. A board cannot just wave a wand and say 'Do it'. As the experiences of Marks & Spencer and Airbus highlight, organizational inertia and organizational fiefdoms get in the way. (It is instructive to read more about these at Airbus, in the reference given at the end of this chapter (Clark, 2006): amid ample signals of disarray, management repeatedly ignored the warning signs.)

- Those linkages take place across multiple dimensions. It is not sufficient for a board to simply say 'product design and the supply chain must work more closely together'. Instead, the business must invest in delivering that closer interaction, working to eliminate the various barriers that we have already seen in Chapter 1 and in the cases of Airbus and Marks & Spencer. These may be:
  - barriers in terms of distance and geography;
  - barriers in terms of organizational structure and fiefdoms;
  - barriers in terms of organizational culture;
  - barriers in terms of relative resource provision;
  - barriers in terms of enterprise systems.

- It is crucial that there is a well-understood and agreed strategy in place in terms of what that linkage is supposed to be optimizing. This strategy – and its clear communication – matters: it is all very well having product design and the supply chain function working closely together, but their dialogues and joint efforts need a focus. At its starkest, the lesson of the Airbus A380 is that while Airbus thought that its strategy was *design for rapid introduction*, a safer strategy might have been *design for manufacture*.

So let's now turn to consider how all this can be put into place within the design-centric business. And in particular, let's start by thinking about the strategy that we want that linkage between product design and the supply chain to actually pursue.

The supply chain begins on the drawing board

A moment's thought reveals that there are many objectives that a design team can jointly pursue in conjunction with their supply chain colleagues.

Is it *design for product cost*, for instance, through such things as clever design, choice of materials and the disciplines of *design for supply* that drive component reuse and product standardization programmes? Or is it *design for rapid time to market* – again employing tactics such as component and supplier reuse, for instance, but also using rapid development methodologies such as concurrent engineering (where such elements as product design, design for manufacture, supplier selection and parts ordering proceed in parallel, or at least in a compressed timeframe)?

Or is it something else altogether – *design for sustainability*, for instance, or *design for manufacture*, or *design for localization* (a technique used by electronics manufacturers with global markets but a highly concentrated manufacturing base)?

Equally, these in turn create further considerations. Design for localization, for instance, may well involve simulation and modelling exercises, in order to balance the number of market-specific product variants produced versus the number of to-be-customized product variants, and to balance both against the resulting inventory holdings and localization centres.

Nor are such issues easily or quickly resolved, as a pivotal paper by Hau Lee, Corey Billington and Brent Carter showed in 1993, examining just such choices in the context of Hewlett-Packard's global desktop printer business. As with optimizing factory and supply chain product flows, modelling and simulation can reduce the time and effort that it takes to resolve these issues, but these can still be significant.

The gains from such efforts, though, are eminently worthwhile, and in Hewlett-Packard's case, were attractive enough for the company to overturn its previous factory-based approach to product localization (the provision of product manuals in the appropriate languages and provision of power cords appropriate for particular geographic markets, etc), and instead carry out localization at regional distribution centres (DCs) located in the Far East and Europe. As Lee, Billington and Carter put it (Lee *et al*, 1993):

> The overall dollar impact is that DC-localization leads to 18 per cent reduction of inventory investment, with no change in its service to customers [off-the-shelf fill-rate]. By continuing the current investment in inventory, Hewlett-Packard could improve off-the-shelf fill-rate by changing from factory- to DC-localisation. Shipping printers to the Far East and European DCs under this alternative [would allow] Hewlett-Packard to reduce transportation cost.

Zara's strategic trade-offs

Localization, though, is only one choice that has to be made (if relevant). Clearly, there are many other potential choices to be made, informed by the business's overall strategy, its product strategy and its supply chain strategy. There are also conflicts to resolve and compromises to be made: design for sustainability, for instance, may involve trade-offs to be made in terms of product cost and choice of supplier. Equally, the time taken in optimizing design for manufacture may impact the eventual time to market.

Zara, the fashion chain retailer owned by Spain's Inditex, is a business that neatly epitomizes how these choices and trade-offs can be exploited to deliver considerable commercial success. Founded in 1975, it is now the world's largest apparel retailer, with brands such as Massimo Dutti, Pull and Bear, Bershka, Stradivarius, Oysho, Zara

Home and Uterqüe. And shortly after the business was founded, Zara's managers began a strategy of rapid innovation in the 1980s, overturning much conventional wisdom.

Rather than aiming to be a *fashion leader*, dictating consumer tastes, Zara instead aimed to be a *fashion follower*, aiming to respond quickly to changing consumer tastes with slick design processes, a vertically integrated manufacturing operation, a just-in-time production process borrowed from the motor industry, a manufacturing practice of maximizing the use of postponement wherever possible (ordering undyed fabric, for instance, and dyeing it only when currently fashionable colours were known), and lean and efficient distribution. As Michael Pich, Ludo Van der Heyden and Nicolas Harle point out (Pich *et al*, 2002):

> Consumers soon learned that there would be something new in Zara every week, and 70 per cent of product range would change fortnightly. Quantities were limited to promote novelty and avoid saturating the market. Instead, successful designs were altered in colour, styling, or accessories. Zara kept up to date with fashion trends without targeting the 'masses'.

But Zara's great genius lay not in its design processes, per se, or its vertical manufacturing operations or its adroit use of postponement. Instead, it lay in the way that it cleverly exploited two aspects of its supply chain – its **downstream supply chain,** in the shape of the stores that it operated, and its **upstream supply chain,** in the shape of its manufacturing and sourcing operations.

It is commonplace these days to talk of Big Data, predictive analytics and mass experimentation. Companies such as Google and Amazon conduct thousands of experiments a year, making subtle changes to their algorithms and websites, and then tweaking those algorithms and websites based on the results. The changes can be so small that individual consumers might never notice them – but en masse, they are significant enough to nudge individuals' collective behaviour in predictable (and profitable) ways.

Long before the term Big Data was coined – decades before – Zara and other Inditex brands mastered the art of seeing stores (and their employees) as valuable sources of marketing intelligence. In a well-known business school case study, professors Kasra Ferdows, José AD Machuca and Michael Lewis describe one such experiment (Ferdows *et al*, 2015):

> When a khaki skirt had sold out in La Coruña after a few hours on the shelves, Isabelle Borges, a product market specialist discovered sales were brisk across other stores. She knew that the skirt was going to be a hit and initiated a process that within days had supplied the skirt to stores on three continents.

A 2012 article in *The New York Times* expanded on the theme, again highlighting how Zara adroitly exploits its downstream supply chain for insights into what is selling well – and what is not selling so well, so as to avoid the kind of over-stocking and subsequent sales and write-downs that we saw impact Marks & Spencer (Hansen, 2012).

> Store managers monitor customers' reactions on what they buy and don't buy, including what they say to sales clerks: 'I like this scooped collar,' or 'I hate zippers at the ankles.' Store managers report this information to headquarters, where it is then transmitted to in-house designers, who develop new designs and send them to factories.

At which point, the *upstream* supply chain kicks in – Zara's ability to quickly factor such information into its design processes, and then turn those designs into finished garments en route to Zara stores. For throughout its history, Zara (and Inditex more generally) has always prized local manufacture over low-cost sourcing from, say, the Far East for the supply chain responsiveness and agility that such local manufacture confers.

In other words, it sources its garments close to its home market – willingly bearing the higher costs of manufacture, in exchange for the shorter lead times and lower levels of pipeline inventory. When fashion tastes change, it can react nimbly – placing on its shelves products that customers do actually want to buy, without being swamped by container-loads of stock that customers don't want and which are destined to be sold at a discount in order to get rid of them.

Zara adroitly blends *local design and production* (for items where responsiveness and agility matter) with **long-distance offshore manufacture** (typically from the Far East) for items where cost (and hence retail price) is important, and responsiveness and agility matter less.

Ferdows, Machuca and Lewis, for instance, describe design teams working in light, spacious, airy rooms at Zara's headquarters in the small Spanish city of La Coruña, physically adjacent to – and working alongside – colleagues responsible for sourcing and production planning, and 'market specialists' – often former store managers – who liaised closely with the managers of the stores for which they were responsible (Ferdows *et al*, 2015).

> The final decision of what, when and how much to produce was normally made by the relevant designer, market specialist, and procurement and production planners. Like their industry peers, these teams worked on next season's designs but, simultaneously and continuously, also updated the current designs.

Speed and responsiveness are further enhanced by a practice of placing over half of the company's manufacturing on a network of factories in Spain, Portugal and North Africa, the majority of which are clustered tightly around its La Coruña headquarters.

A policy of having the most market-sensitive and fashion-centric items manufactured closest to the La Coruña headquarters (from where global distribution also takes place) means that the production process for these items – from start to finish – takes only two to three weeks, less than half the time that might be regarded as an industry norm for domestic production. The balance of production is sourced from suppliers, of which the majority have historically been in Europe, although Far Eastern suppliers in economies such as China, Bangladesh and Vietnam are also used.

In short, it is easy to see why Inditex – which still maintains a long-held policy of not advertising – has come to dominate the so-called 'fast fashion' niche with its Zara brand. Simply put, Zara is incredibly market-sensitive: it looks at what is selling well, lets that information inform its decisions about new designs and has a close-to-home supply chain that can go from design to store in three weeks or even less.

Put another way, by keeping production of the most market-sensitive and fashion-centric products geographically close to its Spanish headquarters and resisting the temptation to source in bulk from low-cost economies in the Far East, the business is consciously trading off product cost against what it sees as the downsides to a Far East-delivered lower product cost – long inventory pipelines, slow responsiveness, the need to discount slow-moving or unwanted items, higher transportation costs and inflexibility. Rather than trying to forecast – months ahead – what customers will actually buy, it simply produces and sells what consumers *are* buying, and what they *do* want to buy.

CASE STUDY   New Look: competing through the design–supply chain interface

Founded in the south-west English town of Taunton by Tom Singh in 1969, New Look had grown by 2008 to become the UK's third-largest women's wear retail brand in the UK, second only to Marks & Spencer and Next.

Positioning itself as a trend-following value fashion retailer, its main competitor within the same market segment is H&M, but other competitors include premium fashion retailers such as River Island, Miss Selfridge, Topshop, Marks & Spencer, Next and Zara, as well as low-price discounters such as Asda Walmart's George, Peacocks, Primark, Matalan, Sainsbury's and Tesco.

Floated on the London Stock Exchange in 1998, by which time it had over 200 stores, including some in Europe, New Look was taken private again in 2006 in partnership with private equity groups Apax Partners and Permira, and in 2015 was sold to South African private equity investment company Brait SA in a deal valued at £780 million.

By then, sales had grown to £1.5 billion, the number of stores exceeded 1,100 and its global reach included Belgium, France, the Netherlands, the Republic of Ireland, Romania, Malta, Malaysia, South Korea, Singapore, Thailand, Indonesia, United Arab Emirates, China, Germany, Russia, Bahrain, Saudi Arabia, Azerbaijan and Poland.

New Look's success is largely down to its ability to adopt a 'fast and flexible' approach to fashion, which is reflected in its ability to quickly transform the latest trends on the fashion catwalk into affordable clothing for the masses. As of 2008, New Look was able to transform a fashion idea into clothing in about 8–12 weeks, with a constant drive to reduce this even further. With new and low-cost styles appearing frequently in store, it is able to keep its stock looking fresh and in tune with the times. Again, as of 2008, New Look's average customer was 29 years old, and visited a New Look store 34 times a year.

New Look originally began with just two designers, who scanned catwalk shows, used trend prediction agencies and created sketches based on their observations. They worked with long lead times, and were physically separate from New Look's buying and merchandising staff.

New Look expanded its team of skilled designers – 25 in number as of 2008 – who are responsible for developing over 50 per cent of designs in house, as well as ensuring that the supply chain can manufacture and deliver those designs in an efficient manner. The designers endeavour to create fashion items, which optimize the balance between the key components of the company's core business concept: fashion, best price and quality.

As New Look's website put it (New Look, 2018):

We put our customers at the heart of everything we do; this helps us to understand how they feel when searching for products – and ultimately, making sure that they feel great when wearing them, whatever the occasion.

To do this, New Look's designers carry out extensive field research and draw inspiration from trade fairs, exhibitions and foreign travel. A major shift in the company's core business process has been creating a closer interface between the design function and the supply chain, and so designers work closely with the company's buyers, pattern makers, merchandisers and – often – suppliers. Deliberately, the team uses standard frameworks and processes, which can be quickly communicated across the supply chain.

A specially developed CAD system makes the design process transparent to their suppliers offshore and has also improved the speed and quality of design samples. In addition, New Look has invested in a Gerber system – a CAD-based technology for pattern cutting, which enables designers and product developers to map out the individual parts for a garment on layers of fabric in order to minimize fabric waste. Again, this technology is shared with suppliers, helping to reduce costs as well as compress the time from design to manufacture.

The company maintains a tight grip on its supply chain and has optimized communication flows in order to ensure a constant exchange of information. In this way, it reckons to be able to respond quickly to changes in fashion trends, order the right products in the appropriate quantities, deliver these products to the correct destinations and finally to sell them at competitive prices.

A large part of New Look's success is due to the speed at which the supply chain can transform its designs into products. This requires a strong input from designers, who must understand the supply chain's capabilities and/or constraints, and also a strong supply chain input to the production process, based on understanding the designer's requirements and then working out how efficiently and effectively designs can be translated into products.

The company sources from many large independent suppliers across Asia and Europe but relies heavily on a small handful of strategic suppliers, and as of 2008 over half of its clothing supplies were sourced from just two suppliers, located in Turkey and China. When new designs were developed, the strategy was to send them out to both these partners, so that manufacture could be initiated in either facility. Then, based on production lead times and volumes, the appropriate decision could be made as to where to manufacture.

As opposed to buying in bulk ahead of a season, New Look's strategy is to make regular purchases throughout the year. This strategy is seen as providing several important benefits. First, it minimizes the amount of capital that is tied up in stocked goods. Second, it reduces the risk of products being left unsold at the end of a season. And third, it minimizes the need for retail markdowns, which have a significant impact on the overall level of profits.

Constrained design versus unconstrained design

Another influential paper on the subject of the linkage between design and the supply chain within design-centric businesses, by veteran supply chain analyst Michael Burkett in 2006, visualized such activities and choices as an evolutionary process, taking place within a 'design for supply maturity model', balancing value against complexity. Simple product design and supply chain considerations deliver limited value, in other words, while more complex considerations deliver higher value. Start with the simple and straightforward, and build from there.

Importantly, though, these various activities and choices imply a genuine two-way dialogue between product designers and supply chain professionals – a dialogue that, in Burkett's terms, marks a higher level of design for supply maturity.

In other words, in contrast to the notion of designing a product and then throwing it over the wall for the manufacturing and supply chain functions to somehow get on with delivering, it is crucial that there is a genuine dialogue to be had, and genuine choices made. Put another way, choices and compromises may emerge that may make product designers uncomfortable or may make supply chain executives uncomfortable – or both. And crucially, the more innovative the product or innovation, the more these conflicts will arise, and require resolution. As Burkett points out (Burkett, 2006):

> While re-use is a sure way to improve volume pricing, it does place a constraint on innovation when applied too rigidly. To design for supply while also developing new products and ensuring long-term profitable growth, companies need to look at the trade-off. What is the benefit of a new innovation versus the cost of supply?

Strategic trade-offs

Burkett's use of the word 'constraint' is revealing. For in order for this dialogue and the resulting trade-offs to take place, the product design function must accept some supply chain-driven limits on its designs and innovations.

Clearly, this won't always be a comfortable (or even acceptable) situation for some designers, and some design functions, within design-centric businesses. In the upper echelons of high fashion *haute couture,* for instance, we can imagine designers being most unwilling to accept supply chain-driven limitations on their designs: from such a designer's perspective of their business model and environment, they may see their role as conceiving designs, and the supply chain's role as taking those designs and manufacturing them, unchanged and without restriction.

That is why it seems sensible to characterize product design functions as either *constrained* or alternatively *unconstrained*. And in this book, we are solely interested in **constrained design** functions as they are the design functions that are willing or able to accept supply chain-specific input from elsewhere in the business in order to inform their design conclusions. Unconstrained design functions are not willing or able to accept that input, and so there is little that a supply chain function can do except attempt to change that unwillingness.

Importantly, however, this is not necessarily the same as saying that constrained design functions must resign themselves to producing designs that are in some sense *compromised* from a design perspective – although those compromises may certainly arise. Instead, the emphasis is on making design decisions while knowing in advance how those decisions will affect the various supply chain cost, risk and lead time outcomes that we have looked at.

Take packaging, for instance. While most (if not all) designers will want to see their products packed in packaging that is functional and aesthetic, and which reflects and supports the relevant brand values, this still opens the door to a wide range of packaging possibilities, all of which might be functional and aesthetic, and which reflect the relevant brand values. But not all of those packaging possibilities will have an equal impact on the supply chain: some will be better than others in terms of their supply chain outcomes.

Must a designer compromise on his or her first packaging concept in order to reflect those superior supply chain outcomes? Some might. But it is to be hoped that the majority of designers will take a pragmatic view and reflect on the fact that what the consumer is *really* buying is the product, not the packaging. In short, it is the product that delivers the utility that the consumer is looking for – and the product will continue to deliver that utility, long after the packaging has been disposed of.

Similarly, with choices involving colour, options, features and materials: yes, designers will have views; and yes, supply chain professionals will have views on those views. Is a departure from a designer's first (or preferred) vision of a product in any of these necessarily a compromise? To some designers, perhaps. But many more, it is to be hoped, will take the pragmatic view that if such a compromise helps to make the product a commercial success, then that compromise will enable their designs to be bought and appreciated by many more consumers than would otherwise be the case.

The 4Cs: aligning product design with the supply chain

So how is this dialogue between the design function and the supply chain to take place within design-centric businesses? How is input from the supply chain best reflected in designers' designs, so as to ensure those optimal supply chain outcomes? Moreover, given the barriers that we have seen to this dialogue and communication, what kind of organization structure helps to bring this dialogue and communication to fruition, so avoiding the kind of disjointed relationship that had a major impact on Airbus and Marks & Spencer?

Think for a few moments about the examples of good – and bad – practice that we have studied so far in Chapters 1 and 2, and it's not difficult to get some pointers. Another clue will have come from the checklists at the end of Chapter 1, with their focus on the nature – and frequency – of the communication between the design function and the supply chain function.

My own research, carried out over many years, points strongly to four critical success factors, each of which can readily be seen in design-centric businesses such as Zara:

- **Cooperation in the extended enterprise,** ensuring that the impact of design decisions are understood by both the supply chain function and external suppliers. Through such cooperation, the business and its suppliers jointly mitigate design–supply chain-related risks and ensure the smooth transition of products through the supply chain to the end customer through early supplier involvement.

- **Co-location of concurrent design teams,** with the design function and the relevant supply chain people being physically close together in order to aid this close cooperation. All the functions that contribute to the design and development of a product are either physically co-located together – or if geographically dispersed, are *virtually* co-located with information transfer on as near a real-time basis as possible – in order to ensure the smooth transition of products from drawing board to market.

- **Cross-functional multidisciplinary teams** of design professionals and supply chain professionals working concurrently, jointly contributing to the design process. These teams can also involve first- and second-tier suppliers, in order to benefit from early supplier involvement.

Figure 2.1 The 4C model for the design-centric business

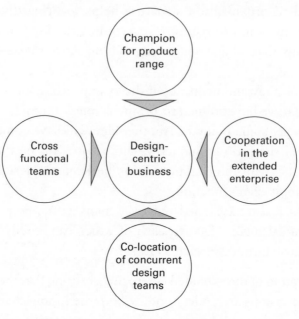

SOURCE Khan and Creazza (2009)

- **A product champion**, tasked with making sure that this dialogue takes place, and who is responsible for making the final call on its outcomes in terms of design decisions. The product champion manages the interface between design and supply chain functions, oversees the concurrent design process, and ensures that there is a match between the product architecture and supply chain design.

Overcoming organizational resistance

None of this is easy, though. As we have seen, one cannot simply wave a magic wand and declare that henceforth, the business will be a design-centric one. In fact, the challenges involved in transforming a traditionally oriented business are significantly greater than those faced by business start-ups set up to be design-centric from day one. It is no accident that companies such as New Look, Apple, Dell and Zara encountered and adopted design-centric organizational principles early in their existence – and also not surprising that the venerable Marks & Spencer (founded in 1884) struggled to embrace design-centric principles.

Overall, indeed, the statistics are gloomy. Overcoming organizational inertia and 'that's the way that we do it here' resistance to change takes time, conviction, strong leadership (and perhaps changes of leadership) and – many times – a 'near-death' experience. It is a fact that many change initiatives either fail or only partially succeed, with research at Henley Forum (nd) and elsewhere repeatedly showing that fewer than 30 per cent of change initiatives count as 'successful'.

Time and again, major enterprises – enterprises with sophisticated management, global or national reach and well-developed internal business processes – announce an intention to change themselves and then proceed to fall flat on their faces.

Consider Microsoft, for instance, which struggled to transform the innovative user-friendly design vision underpinning its Vista and Windows 8 operating systems into products that consumers would actually like, and which businesses would actually want to adopt. Or alternatively, consider media giant News Corporation, which in 2005 spent US $580 million acquiring MySpace in pursuit of a grand 'go digital' change transformation vision, only to eventually sell it for US $34 million.

All of which serves to underscore the fact that the organizational challenges involved in becoming a design-centric business should not be taken lightly. This book is not the place for a detailed exposition on how best to overcome organizational inertia and resistance to change – there are countless books and academic articles on that topic, and the list grows daily.

Many, though, would point to the writings and influential thinking of Harvard Business School's John Kotter and the eight-step model outlined in his 1996 book *Leading Change*. Based on an article entitled *Leading Change: Why Transformation Efforts Fail*, published the previous year in the Harvard Business Review (and which quickly became the journal's best-selling reprint), *Leading Change* provided managers with a whole new language to discuss, describe and deliver change (Kotter, 2012).

Stripped to its essentials, Kotter's model does two things. First, it outlines a series of steps deemed necessary if major change is to take place. And second, it places these in an order of precedence. From the perspective of an organization wanting to deliver major change – and, moreover, to make that change sustainable – the power of Kotter's approach is that it is, almost literally, a checklist. Revised slightly in 2014, Kotter's (by now famous) eight steps have undeniably stood the test of time. While there are other approaches to bringing about organizational transformation, Kotter's eight steps are as good a place to start as any (Kotter, 2012):

1 Establish a sense of urgency.

2 Create a guiding coalition.

3 Develop a vision of a post-change state.

4 Communicate this change vision.

5 Empower others to act on the vision.

6 Generate short-term 'wins'.

7 Consolidate improvements, and produce still more change.

8 Institutionalize, or 'anchor', the new approaches.

As we will see in Chapter 6, Kotter's change management model and methodology is these days one of many. But for a business moving to become more design-centric and uncertain as to how to achieve this – and importantly, how to achieve this as a permanent, lasting change to the organization and its business model – Kotter and similar approaches to change management usefully underscore the fact that mere exhortation is not enough. More than a memo from the chief executive is required.

Product life cycle management (PLM)

As we have seen, one way of building a design-centric business is the '4Cs' approach, which sets out the guiding organizational principles necessary to optimally align product design with the supply chain. As a reminder, these are:

- cooperation in the extended enterprise;
- co-location of concurrent design teams;
- cross-functional multidisciplinary teams;
- a product champion, orchestrating the dialogue.

A technology known as product life cycle management (PLM) aims to help businesses with the first three of these steps, in the form of a PLM system linking together all the relevant parties in bringing products to market and coordinating their activities, in much the same way that enterprise resource planning (ERP) systems do the same for the order-to-invoice process.

In doing so, some might think that the PLM becomes the fourth of these '4C' steps – the product champion orchestrating the dialogue. That view is mistaken and helps to explain instances where businesses have invested in PLM systems, but have still not seen the design-centric transformation that they sought.

PLM is not a substitute for a product champion, but instead provides that product champion with an enhanced tool set so as to coordinate and orchestrate more effectively. More particularly, PLM can also be thought of as a way to get, say, most of the advantages of physical co-location without that physical co-location actually taking place. Instead, the business aims to become design-centric through virtual co-location, rather than physical co-location.

At its simplest, PLM software helps the would-be design-centric business in three key respects:

- **PLM software streamlines the workflow** that takes as its starting point a computer-aided design-derived (CAD) product design or component design, and then builds a complementary set of product or component data and attributes such as material specification, performance requirements, colour, grade and so on – the sort of supply chain-specific data that will be required when going out to sourcing or packaging design. PLM can therefore be seen as a marriage of CAD and product data management – even if these two processes take place in physically separate locations.

- PLM software acts as a single 'source of the truth', or data repository, of all a business's products and components. Again, this helps to transcend the difficulties imposed by distance if co-location is not possible. It is then easy for design functions to access and reuse other designs drawn from outside their own particular immediate design knowledge and responsibilities. A designer working on a new car or home appliance, for instance, can quickly call up and reuse not just the CAD drawing of a component already in existence, but also the associated product data in terms of specifications and attributes.

- PLM software helps a business to capture – and act upon – data from the extended supply chain and the extended enterprise, by providing a means of capturing and integrating product-related quality and performance data from right across the supply chain, such as reports from field service technicians, suppliers, customers and the factory floor. Better still, it can then provide a means of prioritizing and driving improvement projects, combining data analysis of newly identified faults with the corrective actions taken to other previously identified faults – powerfully leveraging engineers' productivity by in effect reusing past knowledge.

CASE STUDY   General Motors

In 1997, General Motors reckoned that it took 48 months to bring a new model to market – four long years during which the cash expended on design, development and prototyping flowed out of the business, without a cent coming in from vehicle sales. But other vehicle manufacturers were quicker.

'At the time, the industry average was between 36 and 40 months – we just weren't competitive,' noted Diane Jurgens, General Motors' Director of Information Systems and Services for Global Product Development Processes, in 2005 (Wheatley, 2005).

As a consequence, not only were competitors recouping their investment in new models faster, but they were also able to more nimbly capitalize on market trends. When a new vehicle catches consumers' imaginations, it's the first few models on the market that earn the best margins: later entrants must offer discounted prices in an attempt to recapture lost market share.

Recognizing that something had to be done, General Motors changed the way that it went about developing new models. In particular, not only were different systems carrying out separate aspects of the product development process, but for the same tasks there were different systems in use internationally. Remarkably, for instance, General Motors had no fewer than 24 different CAD systems in use.

Such diversity simply got in the way of efficiency. A component in use on a Vauxhall vehicle, for example, might be capable of being reused in a new Pontiac or Buick, a reuse that would not only reduce development costs, but also reduce development *time*, and possibly also lower manufacturing costs, thanks to increased economies of scale. Yet with different systems, exploiting such potential was difficult. As a result, General Motors took a decision to standardize on a common set of design tools.

'With a single CAD system, a single product data management system and a single visualization tool, we could share information more readily, and collaborate more effectively,' said Jurgens. And collaborate more effectively with supply chain partners and suppliers, too – who would no longer need to operate multiple systems in order to work with different parts of General Motors.

That collaboration capability was significantly enhanced by another decision: taking advantage of PLM, in this case embodied in a technology solution called Teamcenter from UGS (now part of Siemens), the CAD and product data management software vendor on which General Motors had standardized. Simply put, PLM could help General Motors to manage its design and development process as a coherent whole, rather than as a set of disparate actions. And when managed as a whole, ran the theory, development costs would decrease, while cycle times shortened.

It was a theory amply borne out in practice, it turned out. By 2005, the vehicle development cycle, which once stood at 48 months, took just 18 months, said Jurgens – and in some cases was just 12 months long. And the company's global development process was more productive, too, churning out 33 per cent more new models than formerly.

Globally in 2004, for example, General Motors launched more new products than ever before: 48 vehicles and 13 powertrains, with the result that approximately 24 per cent of General Motors' portfolio volume were 2004 models – basically one out of every four vehicles General Motors sold.

'PLM has produced around a billion dollars in development cost savings, and over US $75 million in measurable material cost savings,' noted Jurgens (Wheatley, 2005).

Checklist
Questions for design professionals

- ☐ Which aspects of your business's supply chain priorities would you change, and why?

- ☐ In your view, which design-centric strategies ('design for manufacture', etc) should your business be pursuing?

- ☐ Is it, in fact, pursuing these strategies, and if not, why not?

- ☐ As a design function, what trade-offs does your business make, eg, product cost versus development time, etc?

- ☐ If the design function was able to collaborate seamlessly with tier-1 suppliers, what do you think it might be possible to achieve?

Checklist
Questions for supply chain professionals

- ☐ If you could change the way that your business's design function operated, what would those changes look like?

- ☐ In your view, which design-centric strategies ('design for manufacture', etc) should your business be pursuing?

- ☐ Is it, in fact, pursuing these strategies, and if not, why not?

- ☐ What prevents your business from emulating Zara's business model?

- ☐ Does your business have a PLM system? And if so, what use does the supply chain function make of it?

References

BBC News (1999) Business: The Company File – Sparks at Marks [Online] available at: http://news.bbc.co.uk/1/hi/business/285252.stm [accessed 12 January 2018]

Burkett, Michael J (2006) Design for Supply: An evolutionary path: design for supply involves more than just part and supplier reuse. Leading companies see benefits from simulation and modelling before and during product launch (Technology), *Supply Chain Management Review*, Peerless Media [retrieved 15 January 2018 from HighBeam Research: https://www.highbeam.com/doc/1G1-155615409.html]

Clark, N (2006) The Airbus Saga: Crossed wires and a multibillion-euro delay – Business – International Herald Tribune, *The New York Times* [Online] available at: http://www.nytimes.com/2006/12/11/business/worldbusiness/11iht-airbus.3860198.html [accessed 12 January 2018]

Ferdows, K, Machuca, J and Lewis, M (2015) *Zara: The World's Largest Fashion Retailer*, case study [Online] available at: https://www.thecasecentre.org/corporate/products/view?id=130606 [accessed 15 January 2018]

Hansen, S (2012) How Zara Grew into the World's Largest Fashion Retailer, *The New York Times* [Online] available at: http://www.nytimes.com/2012/11/11/magazine/how-zara-grew-into-the-worlds-largest-fashion-retailer.html [accessed 12 January 2018]

Henley Forum (nd) Knowledge in Action Issue 32 [Online] available at: https://www.henley.ac.uk/research/research-centres/the-henley-forum-for-organisational-learning-and-knowledge-strategies/folks-knowledge-in-action [accessed 12 January 2018]

Jackson-Proes, A (2001) Trouble deepens as M&S sales decline, *The Telegraph* [Online] available at: http://www.telegraph.co.uk/finance/2725809/Trouble-deepens-as-Marks-and-Spencer-sales-decline.html [accessed 12 January 2018]

Kotter, J (2012) *Leading Change*, Harvard Business Review Press, Brighton, MA

Lee, H, Billington, C and Carter, B (1993) Hewlett-Packard Gains Control of Inventory and Service through Design for Localization, *Interfaces*, **23** (4), pp 1–11

New Look (2018) [Online] available at: http://www.newlookgroup.com/who-we-are/our-brand [accessed 15 January 2018]

Pich, MT, Van Der Heyden, L and Harle, L (2002) *Marks & Spencer and Zara: Process Competition in the Textile Apparel Industry* [Online] available at: https://hbr.org/product/marks-spencer-and-zara-process-competition-in-the-textile-apparel-industry/INS849-PDF-ENG [accessed 12 January 2018]

The Economist (2001) Marks's failed revolution [Online] available at: http://www.economist.com/node/569731 [accessed 11 January 2018]

Wheatley, M (1998) Pile It Low, Sell It Fast, *Management Today*, 1 February, p 62

Wheatley, M (1999) Compaq vs Dell, *Logistics Europe*, September, p 18

Wheatley, M (2005) Boosting productivity, *The Manufacturer* [Online] available at: https://www.themanufacturer.com/author/malcolm-wheatley/ [accessed 15 January 2018]

Further reading

Cargille, B and Fry, C (2006) Design for Supply Chain: Spreading the word across HP [Online] available at: https://kelley.iu.edu/mabert/e730/HP-Design_SCMR_2006.pdf [accessed 15 January 2018]

Johnson, M Eric (2002) Product Design Collaboration: Capturing lost supply chain value in the apparel industry (April), Tuck School of Business Working Paper No. 02-08 [Online] available at SSRN: https://ssrn.com/abstract=307461 or http://dx.doi.org/10.2139/ssrn.307461 [accessed 15 January 2018]

Murphy, J (2008) What a Bright Idea: Innovation Stems from Convergence of Design, Supply Chain, *Supply Chain Brain* [Online] available at: http://www.supplychainbrain.com/index.php?id=7098&type=98&tx_ttnews%5Btt_news%5D=3909&cHash=da03e20e36 [accessed 15 January 2018]

Siemens PLM Software (2016) *Product Lifecycle Management Software* [Online] available at: https://www.plm.automation.siemens.com/en_us/plm/ [accessed 15 January 2018]

Product design and supply chain risk 03

Introduction

Supply chains contain risks, not least risks in terms of costs, complexity and **continuity of supply**. Design decisions can have a major impact on those risks, with unconstrained design processes exposing the business to supply chain disruption, additional costs and inventory holdings, and technology problems.

This chapter explores a number of these risks – principally supplier-related risk, technology-related risk, component or material-related risk, design-related risk and complexity-related risk – and outlines ways in which better cooperation between the design and supply chain functions can reduce them.

Lessons in resilience and complexity

On 11 March 2011, an earthquake measured at 9.0 on the Richter scale hit Japan's north-east. Homes, schools, factories, power lines, roads and communications links were damaged or destroyed. But worse, the earthquake triggered a tsunami, and with awesome power, a series of waves picked up people, vehicles and buildings, and swept them far inland in a maelstrom of devastation that left over 25,000 Japanese citizens missing or dead.

Within hours, the finely honed supply chains of Japanese car manufacturers such as Toyota, Suzuki and Nissan were hit by parts shortages. In the days that followed, parts shortages spread further disruption far beyond Japan's shores, affecting vehicle manufacturers as diverse as Ford, Volvo, General Motors, Renault, Chrysler and PSA Peugeot-Citroën (Wheatley, 2011).

From electronics equipment to paint and from engines to gearboxes, huge numbers of automotive components turned out to be sourced from Japan and its intricate supply chains. Worse, many of these were single-sourced – sometimes not just to a single company, but to a single plant within that company.

A case in point: a metallic paint pigment called Xirallic that was produced at only one factory in the world, owned by German-owned manufacturer Merck KGaA, located in the Japanese coastal town of Onahama and used by manufacturers as diverse as Chrysler, Toyota, General Motors, Ford and Volvo.

Chrysler was forced to restrict orders on vehicles in 10 colours, including the imaginatively named Bronze Star, Rugged Brown, Hunter Green, Ivory and Billet Silver, reported Reuters (Seetharaman, 2011). Ford slowed the production of vehicles in a colour called Tuxedo Black, and also in three variations of red and told dealers that they would not be able to order black-painted Ford Expeditions, Ford Navigators, Ford F-150 pickup trucks and Ford Super Duty pickups.

Ford spokesman Todd Nissen told Reuters that Ford was looking at other materials, to see if they could produce the same shiny effect as Xirallic, and that Ford was also working with Merck to see if the pigment could be produced elsewhere. Merck, meanwhile, wasn't optimistic, explaining that it would be difficult to transfer production to another plant, and that after repairs, customers could expect that it would take between four and eight weeks to resume production.

In the event, it was to be 8 May before the plant resumed production, eight weeks after the earthquake and ensuing tsunami. And no longer is Xirallic sole-sourced: these days, a Merck spokesperson told Reuters in 2016, five years after the disaster, the company now keeps 'multi-month' stocks of Xirallic at warehouses in Japan and other regions around the world. In addition, a second production line was opened up in Germany in 2012.

For another – and completely different – slant on product design-induced risk, consider Fortune 500 appliance manufacturer Whirlpool Corporation. In 2007, the company realized that its battle to deliver ongoing procurement savings was being stymied by an implacable adversary: complexity.

Partly through design proliferation arising through normal organic growth and partly as a result of acquisitions, the company was finding that too many purchased parts and materials turned out to be unique – unique either to particular end products, or to Whirlpool. As a consequence, procurement-led cost reductions were frustratingly difficult to deliver.

In terms of simple components such as switches and water valves, for instance, the range of purchasable part numbers might easily extend into the hundreds. Instead of routinely looking to reuse components that were already in use – and thus already being purchased for other products in the company's product range – Whirlpool management realized that product designers were all too readily reaching for their drawing boards or parts catalogues.

Yet in theory, increased standardization was eminently possible: many components and materials were very similar, but with minor unique characteristics that generally were not adding meaningful value. The more those minor characteristics could be eliminated, ran the logic, the greater the standardization that could be delivered. There were, for instance, 150 different water valves for dishwashers: the goal was to reduce this to 40 or 50 valves (Wheatley, 2010).

Accordingly, Whirlpool set up what it called 'component architecture management' teams, targeted on rationalizing component use. Comprising representatives from its procurement, engineering, technology and product design functions – as well as suppliers – each team was given 16 weeks to conduct a review of the parts that it had been tasked with rationalizing.

Moreover, the goal was not to carry out a 'one off' rationalization exercise. Instead, and significantly, a culture change was also targeted, involving creating an internal culture where the very need for part proliferation was questioned and challenged. And whereas short-term cost reduction through rationalization was intended to deliver 30 per cent of the component standardization initiative, 70 per cent of the initiative's overall ROI was to come from the longer-term benefits of a design process that consciously sought to clamp down on part proliferation, aiming to reuse existing parts and materials, ideally sourced from reliable and trusted suppliers with whom Whirlpool had already established a relationship.

'Longer term, we could see our benefits being driven through continued growth of our strategic suppliers, engineering design changes and the standardization of components, inventory reduction, component and supplier globalization, reduced product changeovers, improved sales and operations planning processes, and supplier migration to low cost country sourcing,' said Whirlpool vice-president of procurement Stefan Grunwald in a press interview (Wheatley, 2010). 'Clearly, we consequently put a greater emphasis and importance on the longer-term benefits.'

Changes were made to the organization structure, he added, giving people clear accountability for component complexity. While the use of a non-standard component such as a water valve was not to be arbitrarily denied, the product designer proposing its use had to provide 'clear and supportable' reasons why a standard part could not be used.

'At the onset [of the initiative], the expected financial improvements were $1 billion annually in direct material savings,' report Robert Rudzki and Robert Trent. 'But this figure does not include the benefits from reduced supply chain complexity, any accounting for cost avoidances, improvements to product quality, or reductions in product development cycle times from having component designs available for reuse. While these types of benefits should be meaningful, they are also difficult to quantify' (Rudzki and Trent, 2011).

That said, by the time the project was reaching its concluding phases, there was little doubt as to the eventual magnitude of such savings. Component count was down 35 per cent, supplier count was down 60 per cent and a direct cost reduction of 7–10 per cent had been delivered. With water valves, there was a 72 per cent reduction in the number of stocked items, a 50 per cent reduction in the number of suppliers and a typical reduction in direct cost of 10–15 per cent. With switches, stocked part numbers fell by 48 per cent, there was a supplier reduction of 66 per cent and a typical cost reduction of 7–10 per cent (Rudzki and Trent, 2011).

Design's impact on the risk profile

Although very different, these two case studies share a common thread. Namely, that – as we have seen – seemingly small and apparently inconsequential decisions taken during product design can go on to have extreme consequences. Specifically, decisions that expose the business to risk.

In the case of automotive designers opting for Merck's Xirallic-enhanced paint colours, the risk was of supply chain disruption, lost sales, disappointed customers and reputational damage. In the case of Whirlpool, the risks were those associated with any episode of stock keeping unit (SKU) or component proliferation, as purchasing leverage dissipated and inventories built. Nor were those risks inconsequential. Whirlpool, remember, expected to save US $1 billion annually in direct material savings – US $1 billion that had built up incrementally over years, one product design decision at a time.

Now, risk is a fact of life, for both businesses and individuals. So by saying that decisions 'expose the business to risk', we do not mean that a situation of *no risk* is replaced by one of *some risk*. Instead, we mean that a given decision has served to increase the risk profile of the business – *adding to the risks that it faces* in other words, rather than reducing those risks.

Put another way, what we are saying is that *good* product design decisions can *reduce* the risks that a business faces and *bad* ones *increase* the risk that a business faces.

This is not merely semantic word-juggling. Instead, it gives us a way of assessing product design in a different way – a way that says the product designer's mission is not merely to design creative and compelling products, but also to do so in a manner that serves to *reduce the risk profile* faced by the business.

Sources of risk

So what, then, *are* these risks? Some we have already seen. And no list of risks can ever be detailed and granular enough to be fully complete: by any measure, a tsunami, for instance, must come fairly low on anyone's list of expectations. But broadly speaking, it is useful to think of the supply chain-related risks that can be impacted by product design decisions as falling into five categories:

- supplier-related risk;
- technology-related risk;
- component- or material-related risk;
- design-related risk;
- complexity-related risk.

Let's look at each in turn.

Supplier-related risk

The broad nature of supplier-related risk is fairly well understood, and most procurement departments have internal methodologies for assessing and mitigating supplier-related risk. These range from the use of supplier qualification procedures, audits and risk-related KPIs as methods of minimizing supplier-related risk, to dual sourcing and contingency inventory as a means of insurance against operations being affected in the event of a risk actually occurring.

Working with a particular supplier, for instance, can expose a business to risks of poor quality control within that supplier; a supplier's financial failure; a supplier's poor internal planning and scheduling, leading to late arrival of materials; and supplier-specific incidents of fire or strike action. Moreover, the choice of a supplier often involves risks associated with where that supplier is located: political instability, logistics infrastructure reliability and quality, and – of course – natural disasters.

Again, Zara is instructive. As we have seen, the use of local manufacture as opposed to, say, sourcing from the Far East serves to increase agility and responsiveness. But it also generally means that a business is dealing with suppliers about which it knows more (because they are nearer), where logistics failures are less likely (port strikes, shipping delays and so on) and where political instability can largely be discounted. On the other hand, while dealing with local suppliers may make the recovery from adverse occurrences quicker and easier, the fact that a supplier is local has little impact on events such as financial failure, fire or quality failures.

Technology-related risk

When making design decisions, product designers make explicit choices about components, materials, and product form and function. But these *explicit* choices also contain *implicit* choices regarding technology – either the technology (or technolog*ies*) embedded in the product that is being designed, the technology required to manufacture that product, or the technology (or technolog*ies*) that are either embedded in the components and material from which it is made up, or which are required to manufacture them.

Sometimes, those technology choices are based on cost considerations: technology A is cheaper than technology B, for instance, even though the end result is broadly the same. Sometimes, those technology choices are based on functionality or capability considerations: by using technology A, for instance, the product can perform this additional function, or carry out that task. Sometimes, too, marketing or aesthetic considerations dictate technologies. And sometimes the technology choice is based on performance, either in terms of reliability, longevity, weight or some other characteristic.

The problem is that these technology choices contain risks – especially when those technologies are new or represent a significant departure from current experience or in-house competencies.

In late 2016, for instance, electronic giant Samsung was forced to permanently discontinue production of its Galaxy Note 7 mobile phone, just two months after launching it. The problem: the phones had shown a propensity for overheating, to the point of catching fire. Keen to beat Apple to the market with a next-generation model, Samsung had rushed the development of the lithium-ion battery that powered the phone. A replacement battery – sourced from a different supplier – omitted the fault that had caused the original battery to overheat but included a different fault that had the same effect.

Sometimes, technology risks pervade a whole industry. In the late 1990s, for instance, electronics manufacturers keen to burnish their sustainability credentials began switching the manufacture of their circuit boards to lead-free solder. From a manufacturing point of view, this resulted in lower yields and teething troubles while manufacturing engineers accustomed themselves to the new technology. But worse was to come: over time, it turned out, lead-free solder had a propensity for growing tin 'whiskers' – hair-like protuberances, composed of tin, which had the potential to cause electronic short-circuits.

By 2005, John Keller, editor in chief at *Military & Aerospace Electronics* was warning that lead-free solder had become a 'train wreck in the making', pointing to its role in a growing number of high-profile failures (Keller, 2005).

Component- or material-related risk

Sometimes, risks are associated not with the technology that is embedded in a specific component or material but from other attributes, generally involving its supply chain or manufacturing characteristics.

In terms of supply chain risks, for instance, it may contain hard-to-get materials with non-standard specifications or peculiarly challenging specifications. Or it may contain materials that are subject to occasional shortages or extreme fluctuations in price – a number of rare earths, for instance, come into this category. Or it may contain materials sourced from regions of the world that are prone to political instability or natural disasters.

In the autumn of 2011, for instance, Thailand was affected by the worst flooding in 50 years, which simultaneously hit seven of the country's largest industrial zones, causing factory output to fall by 36 per cent. In the country's Khlong Luang industrial zone, factory floors lay under up to two metres of polluted water for several weeks, and of the 227 factories operating in the zone, a mere 15 per cent had restarted production six months later. Two of the world's largest manufacturers of computer hard drives – Seagate and Western Digital – had factories in Thailand, and Western Digital reported manufacturing 60 per cent of its hard drives in the country.

'Over the weekend, workers in bright orange life jackets salvaged what they could from the top floors of Western Digital, a single facility producing one quarter of the world's supply of sliders. The ground floor resembled an aquarium, and the loading bays were home to jumping fish' (Fuller, 2011).

Finally, in terms of manufacturing characteristics associated with risk, the specifications associated with specific materials or products may prompt their manufacturers to produce them only in limited quantities, or to fixed schedules. Rationing or 'allocation' may take place in the case of shortages, with limited substitute stocks being available in the marketplace. Some parts or materials, due to their specifications or attributes, may only be available 'to order', on long lead times.

Design-related risk

Design-related dangers don't just lurk in the supply chain: sometimes they are attached to the actual process of design but go on to affect the supply chain. In other words, design errors, and how the process of design is carried out, have ramifications for the supply chain.

Consider the example of the Airbus A380, which we looked at in the previous chapter. The chaos in the supply chain is undoubted, as Airbus struggled to connect together aircraft assemblies containing wires that were too short. But the prime reason for the difficulties lay not in the supply chain – and in the sort of issues that we have reviewed in the last few pages – but in the actual design process, with different CAD design systems in use, different and widely distributed design groups using those different design systems, and a cross-border manufacturing process.

Thankfully, such issues are relatively rare. But what are much more common are design-related risks caused by how designs are disseminated along the supply chain. Consider a company with a 3D CAD system, capable of producing beautifully accurate digital designs, and the ability to apply where necessary simulation and modelling tools to 'virtually test' those designs in close to real time.

So far, so good. But can its suppliers – first tier, second tier and third tier – read and work with those designs? Especially in less well-developed countries, where lower technology is extant, or in industries where paper-based designs have long held sway?

Not surprisingly, horror stories abound of confusion created at the point where digital designs have cascaded down the supply chain to a point where they cannot be read and interpreted. Designs printed out and faxed, for instance; or printed out, sent by mail or courier – and then *redrawn by the receiving business*, using their own technology and conventions. The result, inevitably, is errors and delays – delays while designs are transmitted and translated, and errors when translations or interpretations are flawed.

In some industries, traditions and culture exacerbate the problems. It's one thing for an engineering graduate to use hi-tech CAD design tools; quite another – it might be imagined – for fashion designers, long-accustomed to graceful drawings of willowy figures, to do so. That said, of course, graceful 'concept' drawings are one thing; detailed size and style drawings, complete with cutting and sewing instructions, and specifications for motif and button placement quite another. But even so, those drawings and instructions and specifications have to be communicated to the supply chain somehow – and again, there is ample scope for error and delay.

Eric Johnson, for instance, of America's Tuck School of Business at Dartmouth College, describes the interactions between Dillard's – a US department store retailer – and its supply chain (Johnson, 2002a):

Dillard's employed staff's role was to send 300–500 faxes a day. For every 6,000+ styles produced a 12-page fax was sent, including the bill of material, sketches, cut and sew instructions. When crunch time came, changes happened over the phone or via email with others in the design team left out of the loop, causing confusion and mistakes.

Clearly, the problem with such a system is obvious: not all suppliers will be equally adept at spotting errors or raising questions about issues that they don't understand. Moreover, in the case of suppliers located half a world away, time zones get in the way, adding further to the delay in resolving these issues. Equally, too, it is reasonable to assume that new suppliers will generally prove to be less adept at working with a business than older, tried-and-tested suppliers – in the nature of things, some kind of learning curve is inevitable.

Finally, think about the consequences of poor communications with the supply chain: longer-than-necessary product launches, slower 'time to volume', higher scrap costs, higher communications costs and lower productivity. Once again, Zara, with its tight reliance on close-at-hand, tried-and-tested suppliers, is seen to be adroitly side-stepping a host of issues that might be entailed when working with unfamiliar or geographically remote suppliers.

Complexity-related risk

As if all these risks weren't enough, let's consider the sorts of risks exemplified by Whirlpool's 'component architecture management' initiative. It is all too easy for designers to conceive of fabrics in yet another colour, or equipment components with some minor tweak to an existing item. But these then call into creation a wholly new SKU, and one which – worse – may wind up being ordered from a wholly new supplier.

The result: complexity. Two SKUs, not one. Twenty SKUs, not ten. A hundred SKUs, not fifty. Each to be ordered, stored, insured, located and picked, counted and managed.

And if that wasn't bad enough, part proliferation eats into economies of scale. Large component manufacturing batches become small component manufacturing batches. Two injection moulding dies, not one. Opportunities for supplier leverage dissipated.

And from the perspective of inventory holding, fewer opportunities to smooth out statistical fluctuations in usage through the lower variance of pooled demands – resulting in higher-than-necessary levels of safety stock and lower-than-optimum levels of flexibility and responsiveness.

Add all these drawbacks together, and the risks are very clear: higher levels of inventory, higher costs, higher raw material and component prices, more physical storage and lower levels of responsiveness. Hence, of course, companies' natural interest in avoiding such risks – and Whirlpool's determination to not just eliminate existing proliferation, but to try to prevent it happening in future.

Supply chain professionals understand the risks of complexity, of course. In many ways, coping with the challenges of complexity is Inventory Management 101 – the foundation on which much else in supply chain management is built. But design professionals often don't understand these risks – or worse, do understand them, but imagine that they are someone else's problem.

Strategies for reducing design risk

So how, then, is design risk to be reduced? The key difficulty is that there are obvious tensions at work. Product design, by its very nature, involves innovation, prompting designers to explore the art of the possible. Talk to a designer at almost any one of the design-led companies that we looked at in Chapter 1, and this view of 'design as innovation' is almost inevitably one that comes across. Innovation, these designers would say, is our job: to come up with products, shapes, colours and internal workings that are different from what has gone before.

And of course, they are right. No one – and certainly not this book – is arguing that innovative designs are bad, or that innovation is to take second place to a policy of extensive design reuse and a future built upon the designs of the past. To see what *that* sort of world would look like, look no further than Soviet-era Russia or India's Hindustan

Ambassador car, modelled on a 1950s Morris Oxford, and which was manufactured, almost unchanged, from 1958 to 2014.

So there is no debate: of course design – and designers – must be innovative. And yet, at the same time, they must not be unconstrained. *Unconstrained design*, as we have already seen, ignores the common-sense realities, risks and cost implications of the supply chain and places innovative design above all else. In this book, we celebrate – and promote – *constrained design*, where these realities, risks and cost implications are taken into account, and sensible decisions and compromises reached to build a balance between unconstrained innovation and a process of totally constrained supply chain-led design.

Likewise, no one is saying that conventional methods of mitigating design risks are wrong – although this book does stress that those methods of mitigation have a cost and are a response to a set of risks that do themselves have a cost. So dual sourcing, backup inventory holding, premium freight, backup 'contingency' sourcing and so on, are all sensible strategies, albeit at a cost.

Even so, these strategies, it should be stressed, do not of themselves *reduce* risk. Instead, they are 'get out of jail' cards: plans to put in place for when things go wrong, and for when in-built design risk results in supply chain disruption. They address the outcomes of risk; they do not impact the risk itself.

As such, this book argues that such strategies are sub-optimal. By all means pursue backup inventory, premium freight and dual sourcing as short-term expediencies. But the *best* response to design risk is to *design it out* – and *not* to put in place safety nets.

Collaborative risk reduction

How, though, is this to be achieved? A key difficulty is that product designers often don't think about risk. Typically, they will both be operating in an unconstrained manner, and simply disregarding the notion of supply chain risk; or operating in a constrained manner, but not recognizing supply chain risk as one of the constraints that they should be taking into consideration.

So there is obviously a process of education that needs to take place. Different companies have tended to approach this in different ways, most seeming to rely on some form of 'education by osmosis' – with designers gradually learning from repeated interaction with their supply chain peers that certain suppliers are less risky than others, and how to evaluate potential design characteristics from a risk assessment point of view.

Better by far – as we saw in the previous chapter – is a truly collaborative and integrated design process where product designers and the supply chain work hand in hand, with the risk aspect of design decisions taken into account at the outset, rather than being managed later in the process thereby slowing time to market and potentially delaying the product launch.

Some writers use the term 'convergence' to describe this, where product designers and their supply chain colleagues work collaboratively on designs simultaneously. As we saw in Chapter 2, this kind of collaboration powerfully impacts both design effectiveness *and* supply chain operations, short-circuiting communications difficulties imposed by distance and organization structure and building a virtual multi-functional design team.

And usefully, such convergence also has a powerful impact on risk – especially when, as we shall shortly see, risk assessment is made an explicit part of the process. And what is especially interesting is the way that such product design–supply chain dialogue affects what we might describe as some of the less obvious design-related supply chain risks.

In other words, collaborative dialogue doesn't just help to prevent the selection of a unique component sourced from a factory sitting on a geological fault line in a flood-prone war zone, but helps to address the risk that a product might be late to market, or overly complex, or contain obsolescent technology or scarce materials. Better still, close collaboration between product designers and the supply chain doesn't just operate to identify and reduce risks – instead, it operates proactively so as to build a design process in which risk is less likely, whether that is the risk of being late to market or any of the other supply chain-related or design-related risks that we have discussed.

In a thoughtful article, citing the experiences of several leading design-centric businesses, writer Jean Murphy puts it this way (Murphy, 2008):

> It's the companies that figure out how to drive integration and collaboration with supplier partners and with the supply chain and engineering teams, all at once, and make it truly a joint development team, that are able to launch innovative products quickly.

Jerry McNerney, Senior Director of Transportation, Distribution and Logistics Solutions at Motorola contends (Murphy, 2008):

> In today's environment, you are going to be at a complete disadvantage if you don't have your supply chain teams as a full participant in the development and the life cycle execution of your products.

One way that Cisco leverages the discipline of designing a value stream and a product at the same time is to do trade-offs during the design phase to react to the market quicker, hence the solution is defined by the supply chain characteristics as much as on design features.

Digital collaboration

Finally, let us return to the power of digitally enhanced collaboration. Product life cycle management (PLM) and digital designs do much to speed time-to-market, as we saw in Chapter 2. And – to the extent that they help to reduce design errors – they also have an impact on risk. But their impact on *supply chain risk* is relatively limited, as long as those digital designs remain trapped within the enterprise. And as we saw earlier, links in the communication chain that rely on paper, and mail, and faxes, are both slower and more error prone. As Eric Johnson puts it (Johnson, 2002b):

> At Liz Claiborne, the organizational change of bringing hundreds of users online with digital design tools was painful. After a five year effort they cut design time by 50 per cent. However, the firm realized that their suppliers were less tech savvy or used different design systems and couldn't profit from the designs.

Early attempts to extend the digital data chain were inevitably clumsy. In our personal lives, many of us will recall struggles with transporting and exchanging floppy disks and USB portable drives, email attachments that exceeded limits or got mislaid and file types that couldn't be read.

The advent of what is now called 'the cloud' has changed all that. From cloud-based storage services such as Google Drive, OneDrive and Dropbox, to cloud-based collaboration and social media tools such as Facebook, Twitter and YouTube, all of us now exchange information via the cloud. The cloud acts as a single, secure and robust repository, and each of us interacts with it either over the web (via a web-browser such as Mozilla Firefox, Google Chrome or Internet Explorer), or by specialist 'client' applications – or indeed, both.

So too with digital data in the supply chain. Cloud-based (or web-based; we use the two terms interchangeably) supply chain management and data exchange collaboration platforms vastly simplify the transmission of digital data along the supply chain, accelerating its pace even as it enhances its robustness and accuracy.

Moreover, these services do more than simply serve as design communication hubs: they provide a means of finding suppliers, securely (and digitally) transacting with them and updating them with order and design changes. As with all technology, a process of evolution and vendor consolidation is inevitable. At the time of writing, though, service providers such as E2open, Kinaxis and GT Nexus dominate the field, along with a number of more specialized providers. For businesses keen to both accelerate time to market and reduce supply chain, cloud-based collaboration platforms have much to offer.

Formal risk minimization methodologies

Finally, let us look at how formal management techniques – in particular, 'stage-gate' reviews and failure mode and effect analysis – can help to reduce design-related supply chain risk. As with the discussion on change management methodologies in Chapter 2, it is not the intention to describe exhaustively how these techniques can be applied. The references at the end of the chapter, however, will serve as a useful starting point.

'Stage-gate' reviews

'Stage-gate' reviews are a technique long used in new product development, popularized by Robert G Cooper and Scott J Edgett in a series of books and articles (Cooper, 2001; Cooper and Edgett, 2006), and commercialized by the Product Development Institute and Stage-Gate Inc. The basic idea is for new product development to be regularly subjected to a series of reviews, each of which must be successfully passed before development can proceed to the next stage. As such, the intention is to kill off 'losers' at an early stage of development, freeing up R&D and design resources so that they can be available to work on 'winners', or on whole new concepts. Otherwise, many firms find that once initiated, projects inexorably proceed through to eventual completion, even though they are likely to result in market failure.

Typically, stage-gating will involve formalized methodologies in the early reviews so as to ensure that a genuine market need really exists and that the intended product meets those market needs; moving on in later reviews to ensure that target price and target costs are being met, that the originally identified market need still exists, that competitor action has not made a material difference to the project's rationale and that product test and launch plans have been properly thought through. The overall effect imposes a number of 'gates' through which projects must sequentially pass prior to launch – with each gate providing the option of cancelling the project or returning to the drawing board to refine the design.

Fairly obviously, the very process of stage-gating serves to reduce risk, by forcing management to address what are in effect checklists at each important stage in a product's development, making sure that due process has been followed and that the investment represented by a new product development project is still on course for an acceptable return.

But critically, stage-gating permits a *second level* of risk reduction too, by making risk assessment and mitigation a formal part of the design and design review processes. In other words, as part of the formal review process through which a project must proceed, risks must have been explicitly identified, and appropriate risk elimination and risk-elimination strategies put in place. A moment's thought reveals that such a process would be invaluable for dealing with many of the risks that we have discussed in this chapter. Where are suppliers located? What dual sourcing opportunities exist? How are designs being communicated? Are existing components being reused wherever possible? And so on, and so on: the questions are explicitly asked, and satisfactory answers must be put forward before development can proceed.

Indeed, some experts advocate a *third level* of risk reduction – consciously revisiting these risk assessments at subsequent stages of the review process, making sure that risk perceptions and mitigation strategies identified at earlier stages of the process still remain valid. As Keith Goffin, Professor of Innovation and New Product Development at Cranfield University's School of Management, puts it (Goffin and Mitchell, 2005):

> As projects evolve, the risks they face evolve too – and the sooner you can get a handle on those risks, the sooner you can decide to kill off projects that no longer fit your risk criteria.

Failure mode and effect analysis (FMEA)

Failure mode and effect analysis (FMEA) – sometimes referred to in the plural as failure Modes and Effects Analysis – is a technique developed by the United States military in the 1940s. As Martin Christopher points out, although frequently associated with Japanese-style total quality management, it is especially applicable to supply chain risk management (Christopher, 2011). Its basic principle is that it is impossible for an organization to devote enough resources to mitigate *all* risks, and so some means of prioritizing mitigation measures is required.

First, the relevant risks must be identified, and their consequences understood. As Christopher (Christopher, 2011) puts it, this involves asking three questions:

- What could go wrong?
- What effect would this failure have?
- What are the key causes of this failure?

Second, these risks – and their mitigation – must be prioritized. Logically, that prioritization is based on:

- the seriousness of the consequences should a given risk in fact occur;
- the likelihood of that risk occurring;
- the likelihood of early detection of the occurrence, and therefore the probability that remedial measures can be quickly effected.

In the literature, and in its classic engineering application, FMEA is generally represented in the form of tables or spreadsheets. But since its adoption within supply chain management, FMEA is often encountered in a simplified, two-dimensional form, using just the first two of these prioritization measures – the seriousness of the consequences of a risk, or its *impact*; and the *likelihood* of that risk occurring. It is then possible to plot specific risks on a chart, or (more usually) a four-quadrant box, showing impact on the vertical axis and likelihood on the horizontal axis, as shown in Figure 3.1.

That said, there is nothing sacrosanct about the four quadrants: it is perfectly possible to use a 3x3 9-quadrant grid (low, medium, high) or a 5x5 25-quadrant grid (very low, low, medium, high, very high).

FMEA therefore provides a very useful framework for assessing and mitigating risks, once they have been identified. At its simplest, it forces management to allocate a level probability to risks, estimate the severity of their consequences and draw up a list of prioritized mitigation strategies. As Paul Dittmann of America's University of Tennessee argues, in a paper highlighting aspects of the approach at work in a supply chain context, this is very powerful (Dittman, 2014):

Figure 3.1 A simple FMEA 'risk matrix'

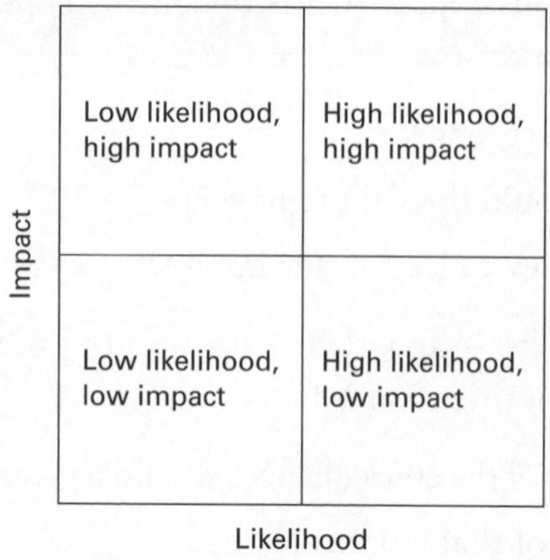

Low likelihood, high impact	High likelihood, high impact
Low likelihood, low impact	High likelihood, low impact

Impact (vertical axis)

Likelihood

SOURCE author

[FMEA] allows [managements] to determine which risks require a mitigation plan, and which are too low impact/unlikely to warrant the effort. Given that risk analysis has a subjective component, reaching consensus is critical. The power of FMEA lies in its use as a framework to debate risks with the supply chain strategy team.

So how does all this work in practice? To see 'stage-gate' reviews and failure mode and effect analysis at work, let us now return – once again! – to Marks & Spencer (Khan and Cutler-Greaves, 2008). In Chapter 2, we looked at how the iconic British retailer's sales and profits had collapsed in the late 1990s, as its lengthy design and production cycles increasingly exposed it to risks that nimbler competitors were able to sidestep. As we saw, the company had been slow to see the potential of globalization and had been caught out by the weather in the winter of 2003 varying significantly from its expectations.

In adopting global sourcing, Marks & Spencer departed from its traditional approach to procurement, specifically adopting an approach called 'direct buying', where buying group teams worked directly with suppliers at the point of manufacture. A business change unit was formed, bringing together the key people and decision makers across the business, in order to help manage the changes that were going to happen and ensure a smooth transition. They were in effect the arbiters of change, any change was reported to them and reviewed against the total business picture.

Clearly, having a robust risk assessment was important in satisfying corporate governance requirements, and providing shareholders and investors with transparency on decisions made. A risk assessment process therefore supported the changes at three levels – board level, programme level and project level – aiming to provide an independent perspective on the risks faced by the business as it engaged with suppliers in a new way.

A series of programme governance guidelines incorporating a risk methodology was introduced, with regular project meetings set up at key stages in the process – the 'stage-gate' process that we have seen but focusing more specifically on risk. Simple tools were used to support this, including a 3x3 FMEA-style risk matrix and risk register, and action plan templates were designed so as to ensure a consistent approach across the business.

First, project teams brainstormed a broad range of risks that were seen as a threat to the achievement of their objective, such as a lack of resource, process failure, external event or opportunities that they might miss. Where appropriate, similar risks were grouped together to produce a more manageable number of risks – perhaps 15 to 25 in total. Then, as in FMEA, the identified risks were examined to determine the impact that the risk would have against achieving the objective, and the likelihood that the given event would occur.

In terms of impact, this was expressed as a percentage of the plan's key financial target set over the period of the plan, or – alternatively – non-financial targets such as brand damage and reputational damage, leading to loss of sales.

The likelihood of an event was expressed as a percentage chance – or probability – of it occurring over the plan period.

Then, a 3x3 FMEA-style risk matrix was determined, and each risk placed in an appropriate cell of the matrix. The risk characterizations in question ranged from 'Very low' (remote possibility,

Figure 3.2   Impact definition criteria

Score	Brand	Description	Impact
3	Critical	Serious; difficult to recover from	>10% brand damage; loss of sales
2	Major	Serious; able to recover from, but with significant effort	2–10% short-term brand damage
1	Manageable	Able to recover from with minor effort	<2% Brand damage; managed internally

SOURCE author

Figure 3.3   Likelihood definition criteria

Score	Brand	Description	Impact
3	Likely	Expected to occur during the period of exposure	>50% probability
2	Possible	Possible, and could occur during the period of exposure	10–50% probability
1	Remote	Possible, but *not* expected to occur during the period of exposure	<10% probability

SOURCE author

Figure 3.4   3x3 risk matrix

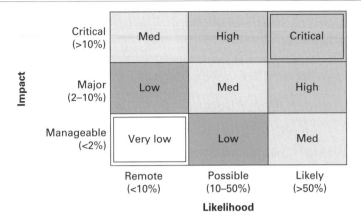

Critical (>10%)	Med	High	Critical
Major (2–10%)	Low	Med	High
Manageable (<2%)	Very low	Low	Med
	Remote (<10%)	Possible (10–50%)	Likely (>50%)

Impact (vertical axis label)

Likelihood

manageable outcome) to 'High' (possible with a critical outcome, or alternatively, likely with a major outcome) to 'Critical' (likely with a critical outcome).

Finally, risks were prioritized by calculating their risk quota – in other words, multiplying the scores for impact and likelihood – which logically results in risk scores of 1 (1x1) to 9 (3x3). For ease, these scores were grouped into high, medium and low-priority risks, with those scoring 6 or 9 ('high' and 'critical') receiving the highest priority attention.

It could be argued that a saving grace for Marks & Spencer was ensuring supply chain risk management as a fundamental process. Assessing risk formally as a continuous process by recognizing that risk evolves and changes throughout the lifetime of a project is key to the governance process, which ensures regular reviews and tracking on progress. Internal audits provide the assurance that risks are mitigated to suitable agreed standards.

Checklist
Questions for design professionals

☐ What supply chain risks have affected your business? Would different design decisions have made a difference – and how?

☐ Is conscious risk-avoidance a part of your regular design reviews and processes?

☐ When specifying parts, components and materials, are you aware of their manufacturing location?

☐ Are you able to interact digitally with all your business's suppliers of direct materials? If not, with what percentage can you interact digitally?

☐ If you have regular joint meetings with supply chain colleagues, does a review of risk form part of these discussions?

Checklist
Questions for supply chain professionals

☐ Are new product designs regularly assessed for supply chain risk before going into production? If not, why not?

☐ Are you aware of the manufacturing locations and inbound logistics routes for all your tier-1 direct materials?

☐ Do you ever talk to tier-1 suppliers about their risk minimization practices?

☐ Could changes to your existing products significantly de-risk their designs? What prevents these changes from taking place?

☐ Can your business interact digitally with all your suppliers to exchange data on designs, product specifications and bills of materials? If not, with what percentage?

References

Christopher, M (2011) *Logistics and Supply Chain Management: Creating value-adding networks*, 4th edn, Financial Times Prentice Hall, London

Cooper, R (2001) *Winning at New Products*, 3rd edn, Perseus Books, Cambridge, MA

Cooper, RG and Edgett, SJ (2006) Stage-Gate® and the Critical Success Factors for New Product Development, *BPTrends*, July [Online] available at: http://www.bptrends.com/publicationfiles/07-06-ART-Stage-GateForProductDev-Cooper-Edgett1.pdf [accessed 19 January 2018]

Dittman, JP (2014) *Managing Risk in the Global Supply Chain: A report by the supply chain management faculty at the University of Tennessee*, The Global Supply Chain Institute, The University of Tennessee College of Business Administration

Fuller, T (2011) Thailand Flooding Cripples Hard-Drive Suppliers, *The New York Times* [Online] available at: http://www.nytimes.com/2011/11/07/business/global/07iht-floods07.html [accessed 19 January 2018]

Goffin, K and Mitchell, R (2005) The Risks of Innovation, *Financial Times*, September 30

Johnson, ME (2002) Product Design Collaboration: Capturing lost supply chain value in the apparel industry, *Textile Digest,* November

Keller, J (2005) Lead-Free Solder: A train wreck in the making, *Military Aerospace* [Online] available at: http://www.militaryaerospace.com/articles/print/volume-16/issue-10/news/trends/lead-free-solder-a-train-wreck-in-the-making.html [accessed 19 January 2018]

Khan, O and Cutler-Greaves, Y (2008) Mitigating Supply Chain Risk through Improved Agility: Lessons from a UK retailer, *International Journal of Agile Systems and Management*, 3 (3–4), pp 263–81

Murphy, J (2008) What a Bright Idea: Innovation stems from convergence of design, supply chain excellence, *Supply Chain Brain* [Online] available at: http://www.supplychainbrain.com/index.php?id=7098&type=98&tx_ttnews%5Btt_news%5D=3909&cHash=da03e20e36_[accessed 19 January 2018]

Rudzki, R and Trent, R (2011) *Next Level Supply Management Excellence*, J Ross Publishing, Fort Lauderdale, FL

Seetharaman, D (2011) *Automakers Face Paint Shortage after Japan Quake*, Reuters [Online] available at: https://www.reuters.com/article/us-japan-pigment/automakers-face-paint-shortage-after-japan-quake-idUSTRE72P04B20110326 [accessed 1 February 2018]

Wheatley, M (2010) Awash with Ideas, *Procurement Leaders*, (24), pp 28–31

Wheatley, M (2011) After the Disaster in Japan, *Automotive Logistics* [Online] available at: http://automotivelogistics.media/intelligence/after-the-disaster [accessed 19 January 2018]

Product design and the pursuit of agility

04

Introduction

In businesses and supply chains, **agility** is a major determinant of competitive edge. But giving customers the products that they want, when they want them, and at a price that they are happy to pay, is no easy matter. Sometimes, for instance, customers may not know what they want.

This chapter looks at the role that the product design process plays in determining agility, and at how *closer links with suppliers and customers* can help. It explores the role played by **concurrent engineering** and looks at how *postponement strategies* can simplify the tricky challenge of trying to predict well in advance exactly what it is that customers want.

Virtual success: the development of the Boeing 777

Boeing's family of 777 airliners stands out for many reasons. At the time of writing – almost a quarter of a century after the aircraft's first flight into the skies above Boeing's Everett assembly plant, just north of Seattle – the 777 is the world's largest twin-engine aircraft, with its best-selling 777-300ER variant capable of flying almost 400 people a distance of 7,370 miles. It was Boeing's first-ever 'fly by wire' aircraft, one of its most successful – with almost 2,000 aircraft ordered as of March 2017 – and with its 777-800 and 777-900 variants due to enter service in 2020, should still be flying almost 50 years after its launch.

To design and supply chain professionals, however, its appeal is more subtle. It was not only the first commercial aircraft to be designed entirely by computer-aided design (CAD) technology, but also serves as a showcase for a team-based approach to design known as concurrent engineering, or simultaneous engineering. Boeing had experimented with the concept before, but this time it was for real: right from the start, decided Boeing management back in 1988, its new airliner – intended to compete with Airbus's A330 and A340 models, and replace airlines' ageing McDonnell Douglas DC-10 and Lockheed L-1011 'Tristar' aircraft – would be designed using concurrent engineering techniques.

Concurrent engineering is perhaps best understood by contrasting it with traditional, so-called 'sequential' approaches to design. Here, products are first conceived, then designed, then marketing people get to look at them, then production engineers start to figure out how to make them, then suppliers have to gear up to produce the components – and finally, at the end of all this, the manufacturing people can start production.

At which point, they might find that the product's parts don't quite fit together, or that its tooling needs reworking, or that the manufacturing process that worked on the development laboratory's bench doesn't quite scale up as anticipated. And so, still more time is lost making corrections. Nor, even when the product has been manufactured, is there any guarantee that it will find favour in the marketplace: customers have not been involved in the development and design process, and in the intervening period between product conception and market launch, customer tastes might have varied, or better competitors' products reached the market.

Instead, the idea behind concurrent engineering is to try to undertake these various steps in parallel. Once the idea for a product has been conceived, the idea is handed over to a special project team, who are then responsible for it until first deliveries commence. The team contains people from all the functions necessary to achieve this: design, marketing, production engineering, manufacturing and purchasing, for instance. Customer input is often sought, too.

So instead of each function completing its aspect of the project before then passing it to another, the team tries to act as a cohesive whole. While the design is being developed, the marketing implications are fed back into the design to produce further refinements. And while this is going on, production engineers and shop-floor people start thinking about how best to manufacture it, working with design engineers to incorporate any ideas they have on how to make it easier to put together. And purchasing staff and engineers work with suppliers to incorporate their ideas as well.

Moreover, this team-based, parallel-working approach to product design not only reduces the timescales taken to bring products to market, it also tends to lead to better products, as the collaborative process throws up opportunities to improve designs so as to reduce costs, improve manufacturability and increase customer appeal. Decisions can be taken quickly, interdepartment conflicts readily resolved and input to the design process sought from the broadest possible set of informed sources – while concurrently delivering a faster, more agile design process and product launch.

A break with tradition

How Boeing came to decide on concurrent engineering as the design basis for the 777 airliner has been well documented, by insiders such as Alan Mulally, for example, the former Chief Executive of Boeing Commercial Airplanes and Vice-President and General Manager of the 777 programme (and later President and Chief Executive of Ford Motor Company); Wolf Glende, Chief Systems Engineer for the 777 programme; and interviews given by 25-year company veteran Ron Ostrowski, Boeing's 777 Director of Engineering.

The problem was simple: like many engineering-led businesses, Boeing had become accustomed to telling its customers what they wanted. It was a practice that had served the company well when on familiar territory, as its best-selling short- to medium-range 737 airliner abundantly demonstrates. But it was riskier when Boeing broke new ground: with development costs estimated at US $1.2 billion over a four-year period, at a time when Boeing itself was capitalized at just US $372 million, Boeing bosses had famously 'bet the company' in developing the 747 'jumbo jet' airliner – and if the aircraft had not been a success, Boeing itself would have foundered (Glende, 1997).

And with the aircraft that was to become the 777, Boeing undeniably *was* breaking new ground. Attempts to interest airline customers with an aircraft that was essentially derived from the existing Boeing 767, and using an identically sized fuselage, had fallen on stony ground. And in addition to demanding a wider fuselage, airlines were making a number of other awkward demands, too.

Boeing's response, decided in a series of high-level meetings, was to conclude that its conventional approach to aircraft design and development was unlikely to do the job. Breaking completely with the past, it decided to design and manufacture the aircraft in an entirely new way.

In doing so, company management undoubtedly had in mind the development of the original 747 airliner, which not only exposed Boeing to significant financial risk, but also took place at considerable speed, with the prototype aircraft rolled out in less than three years after Boeing and lead customer Pan-American Airways had signed a letter of intent.

Coming in overweight and underpowered, the project had employed 4,500 engineers at its peak – yet even so, had required reassigning some of Boeing's top engineering talent to the urgent task of trimming excess weight off the design, not least in order to enable the aircraft to safely take off. At least some of the blame for both problems, reckoned knowledgeable observers, lay squarely in the supply chain: engine manufacturer Pratt & Whitney had struggled to meet the thrust targets that each of the aircraft's four JT-9D engines had to deliver, while Pan-American's grandiose ambitions for the aircraft had contributed to the excess weight (Glende, 1997).

The result was that while the initial variant of the 747, the 747-100, proved adequate for North American domestic routes, its suitability for long international routes was limited. As a workaround, it wasn't long before Boeing began building the 747SP, a variant that overcame this limitation by shortening the aircraft's fuselage by 48 feet, thus eliminating weight and extending its range. It took until 1971 before the first 700-200 went into service, which finally gave the 747 the range and payload required for long oceanic routes. Hugely popular, and in service for many years, the 747-200 is arguably the aircraft that the original 747-100 should have been.

Listen to the customer

To begin with, Boeing management recognized that enticing satisfied customers away from McDonnell and Airbus's established wide-body offerings called for the company to listen rather than talk. Consequently, it made an unprecedented effort to solicit customer input into the design and development process and understand not only the type of aircraft that customers wanted, but also how they wanted to use it and maintain it. 'We had to learn to think like an airline: we knew how to build aircraft, but not how to operate them,' is how former 777 Director of Engineering Ostrowski put it.

Eight 'technically strong' airlines were selected – including operators such as British Airways, United Airlines, Japan Airlines and Qantas – and asked by Boeing to start working with it on specifying and designing the new aircraft. It was a bold step: these were companies that were more used to competing with each other than collaborating, and professional facilitators were employed to encourage people to speak frankly.

Each airline was also tasked with representing its geographic region, with British Airways charged with making sure that its experience and understanding of the European market was reflected in the design. Boeing did not want to have to subsequently redesign a North American-inspired aircraft in order to make it meet the demands of European airlines.

Consensus soon emerged – uncomfortably so, as it transpired. Boeing's broad intention had been to design the new aircraft as a 'stretched' 767; thereby economically reusing an already developed fuselage. Instead, the airlines wanted an aircraft 25 per cent wider – and got it, making the 777 only 10 inches narrower than the 747.

Another surprise for Boeing was the airlines' insistence on interior flexibility. On long-haul flights, galleys and toilets are traditionally used as the dividers between first class, business class and economy. This means freezing the proportion of each aircraft allocated to each class at the aircraft ordering stage – despite knowing that the actual requirement might be quite different 5 or 10 years into an aircraft's life, necessitating an expensive and semi-permanent refit.

Moreover, the split between classes often varies according to route, and changes again over time. Every passenger forced to fly economy instead of business class, or business class instead of first class, represented a significant loss of revenue over the operating life of the aircraft. The need was clear, but aircraft lavatories and galleys are solid constructions, plumbed into the aircraft's power and water systems. To customers, though, the challenge was simple: why couldn't Boeing design galleys and toilets – together with their plumbing and power systems – that slid up and down the fuselage?

Another sore point was service readiness. The last new aircraft that Boeing had launched, the stretched 747-400, had needed 300 extra engineers assigned to it after the aircraft had entered service, in order to sort out bugs that hadn't been spotted during design or manufacture. Again, the rationale was obvious: once in service, aircraft profitability is driven by utilization, and hours on the ground being repaired have a high opportunity cost – not to mention an impact on customer satisfaction and on-time departure metrics. The goal: an aircraft that not only really *was* ready for service, when finally released into commercial service, but which was also easier to maintain.

Having given the airlines a blank sheet of paper on which to specify their ideal aircraft, Boeing had no choice but to accept the resulting design. Yet the upshot was clear: what had started out as a project to stretch the 767 had instead turned into something with all the hallmarks of one of the massive gambles of the past. Namely, an all-new design for the largest twin-engine aircraft in the world, 25 per cent wider than the 767, incorporating a radical degree of interior flexibility, and with an unprecedented level of service reliability – the latter reportedly enshrined in a handwritten promise to United Airlines, who had by now ordered 34 of the aircraft, with options on a second 34. Boeing's problem: how were they going to meet this challenge?

New ways of working

The second plank in Boeing's strategy was once again to break with the past – this time in terms of the interface between its design and assembly functions. Rather than design an aircraft and then hand the design over to production for assembly, the two functions would work closely together from the start.

Several hundred 'design/build' teams, in Boeing terminology, were accordingly set up to take total design responsibility for the 777. Alongside the engineers who would normally design the aircraft were assembly workers advising how to make the various parts fit together more efficiently, mechanics helping to make them easier to maintain, tooling experts designing jigs and equipment, and – once again – customers.

The problem was the scale. While concurrent engineering wasn't new, instances where it had been applied on products of this level of complexity were rare. Boeing would have thousands of people working on the 777, with the aircraft itself containing 3 million parts. Could concurrent engineering be made to work on this scale?

The most obvious challenge: Boeing's internal culture. Quite simply, Boeing had never built an aircraft in this manner before, so none of these groups had worked with each other. This is a difficulty encountered by every business implementing concurrent engineering, of course, but was seen as a particular difficulty in the rigidly hierarchical Boeing culture. '[It was] probably the biggest challenge that we faced,' explained Chief Engineer Henry Shomber, with Boeing since 1958, and who led a number of communications and culture-changing sessions to try to break down walls and barriers that had been built up over the company's 75-year history.

But putting teams of people together in the same room to design aircraft components – engineers, assembly workers, maintenance people, customer representatives – was only the start. It was clear that the new process, given the extent of cross-referral and consultation involved, would be very slow – and yet the company needed to get the 777 into the marketplace quickly. Just as importantly, how could each team ensure that its component fitted together with other teams' to make a whole aircraft?

Again eschewing conventional terminology, Boeing decided to discard its traditional drawing-board-based approach to the design process and instead employ 'digital definition' to design each part. Once designed, these parts would then be integrated with each other by 'digital pre-assembly' – essentially building the entire aircraft inside a computer first.

Today, when simulation is much more commonplace, such a step no longer seems so revolutionary. Indeed, given the move from 2D designs to 3D designs, created on powerful CAD workstations, integrating those designs together in a form of digital assembly makes logical sense. Even so, the computing workload is high and often acts as a barrier. Back in the early 1990s, the challenge was proportionately greater, of course. Nevertheless, Boeing management was adamant: the 777 was to be the first jet airliner to be 100 per cent digitally designed using three-dimensional solids technology and then electronically pre-assembled, avoiding the need for costly, full scale mock-ups.

Consequently, Boeing launched a worldwide search for the best technology to do this and found it in Europe, in the form of CATIA, an advanced piece of CAD software developed in France by Dassault Systèmes. This enabled Boeing to design the aircraft 100 per cent electronically, with some 2,200 computer workstations linked to what was then reported to be the largest cluster of IBM mainframes in the world.

But the investment was worth it: for the first time, Boeing could 'build' an aircraft inside a computer, before a single piece of metal had been machined. This not only enabled it to make sure that everything fitted together efficiently and without interference clashes – a common bugbear when fitting together 2D designs in the physical world – but just as importantly, help Boeing to meet its customers' goal of improved servicing and reliability. It was possible, for example, to model a human mechanic, dubbed 'CATIA-man' by Boeing engineers, who was able to show that a real-life human mechanic wouldn't be tall enough to change the bulb in the red navigation light on top of the aircraft.

Agility through concurrent engineering

The history of the development of the Boeing 777 is a story with many lessons, and on many levels. And as we saw in Chapter 2, with the difficulties encountered by Airbus with its Airbus A380 aircraft, not all of those lessons had been absorbed, even 30 or so years later.

And yet perhaps the most surprising thing about some of those lessons – the value of concurrent engineering in compressing development cycle times, improving product marketability and design quality, and the advantages of engaging early with the supply chain in the form of suppliers and customers, for instance – is that they took so long to be learned, and appeared at the outset so revolutionary. Even with the most rudimentary knowledge of critical path analysis, project planning or business process design, it is not difficult to see that performing activities concurrently, rather than in sequence, will compress the overall timescale.

That is why, in the '4C' model introduced in Chapter 2, we speak not just of the co-location of engineering and design teams, but instead of the co-location of *concurrent* engineering and design teams. And yet, for most of the modern industrial era, the logic of concurrency has been deemed to be not applicable to product design.

As with Henry Ford's Model T, or Harry Ferguson's post-war T20 tractor, the prevailing logic has been to design a product – often making sweeping assumptions about what customers would want – and then hand that design over to manufacturing and the supply chain. Once produced, an army of marketing people would then try to persuade customers that they wanted it, despite its inadequacies and flaws. Perhaps not surprisingly, for every Ford Model T, the occasional failure of this way of working was inevitable – with Ford again producing one of the poster children of such failure in the shape of the disastrous Ford Edsel. And if soliciting structured and considering downstream supply chain input, in the form of customer requirements, has been relatively rare, upstream input – from suppliers – has been seen as even more revolutionary.

The pursuit of agility

The driving force that finally impelled businesses to break with this way of working is not difficult to determine. For most of the 20th century, industrial companies had pursued a strategy of mass production, aiming to win market share by driving down product costs and increasing products' affordability for the newly emerging middle class.

Cars, home appliances, radios and televisions – whatever the product category, features and user-selectable options played second fiddle to price and raw functionality. For a family that had hitherto only been able to afford bus travel or a motorcycle, a car – any car – was a huge step forward. Likewise, a family's first refrigerator, freezer or television. A rich feature set was secondary, a reality perhaps best characterized by Henry Ford's famous remark about customers being able to choose any colour they liked, as long as it was black.

But gradually, in the post-war years, that began to change. And by the 1970s, product life cycles were accelerating, and product innovation began to focus less on raw capabilities, and more on differentiation through feature sets and attributes. As such, it dovetailed well with a complementary strategy of *competing through agility* – recognizing that increasingly affluent consumers wanted not simply affordable and more feature-rich consumer durables and other products, but also a greater variety, products that were available more quickly, and in more formats, and which had a greater assurance of availability.

Consumers were no longer prepared to wait for stocks to become available (or products manufactured to order), in other words, but instead wanted them *now*. Agility – the ability to respond quickly and flexibly to changing consumer demands, tastes and fashions – began to supplant simple price competition as a business strategy.

Lee characterizes agility like this (Lee, 2004):

> Agile supply chains respond to uncertainties in a rapid, flexible, cost-effective and reliable manner. Building agility requires strong supplier relationships, the right buffer inventory, appropriate capacity levels, product and process design with postponement...

From the perspective of this book and its focus on the links between product design and the supply chain, three things stand out from this characterization of agility.

First, as we will see later in this chapter, products and business processes should be designed so that they are able to exploit *postponement strategies* to play a valuable part in contributing to a business's – and a supply chain's – agility. We will come later to a more detailed look at postponement but suffice to say at this point in the chapter that postponement involves retaining the maximum possible element of flexibility, by making it possible to construct products, manufacturing processes and logistics processes so that the point at which a product is 'frozen' into a particular specification or SKU occurs as close as possible to individual customer demand.

Second, Lee writes of *strong supplier relationships* as contributing to agility. Clearly, strong supplier relationships do this by making it possible to vary output levels, inventory levels and lead times so as to flexibly meet and satisfy as much customer demand as possible. But – in the context of this book – those strong supplier relationships also play another pivotal role: helping to contribute to the design process itself, by harnessing a supplier's (presumably greater) knowledge of its own core products and technologies, so as to add that insight to the customer's own design.

While not explicitly mentioned by Lee, those strong supplier relationships have a reciprocal dimension, or mirror image: *strong customer relationships*. As with strong supplier relationships, strong customer relationships contribute towards agility by providing a more precise understanding of real underlying customer demands than could be obtained otherwise. As Christopher points out, 'a common failing in business is to assume that "we know what our customers want"' (Christopher, 1998). That assumption is often mistaken, and – as with strong supplier relationships – it is usually better to make the effort to open up a dialogue directly with the customer. As we saw earlier in this chapter, that was precisely the approach taken by Boeing when developing the 777 when the company recognized that it had become overly reliant on telling its customers what they wanted.

And as we saw with Boeing and the 777, a powerful way to leverage strong customer and supplier relationships in the context of product design is to do so through concurrent engineering.

Concurrent engineering

The term 'concurrent engineering' can be traced back to a seminal 1988 report by Robert Winner *et al* into the development of weapons systems. In fairness, Winner *et al* are not usually regarded as inventing the concept, merely defining and describing it. As Winner concedes, concurrent engineering 'is a common sense approach to product development, which has been known for some time, [although] modern techniques have facilitated its use.' (Winner *et al*, 1988).

Also known – as we saw at the beginning of this chapter – as 'simultaneous engineering', it dates back much earlier than the 1980s and can be seen at work – albeit unnamed – in the rapid development of aircraft in World War II. But Winner *et al* not only usefully defined concurrent engineering, but also outlined the cultural and managerial changes necessary for its successful adoption. First, the definition (Winner *et al*, 1988):

Concurrent engineering is a systematic approach to the integrated, concurrent design of products and their related processes, including manufacture and support. This approach is intended to cause the developers, from the outset, to consider all elements of the product life cycle from conception through to disposal, including quality, cost, schedule and user requirements.

And, as Winner *et al* noted, what studies that there had been into the effectiveness of concurrent engineering reported some attractive benefits:

- Improvements in design quality, achieving reductions of over 50 per cent in the number of engineering change orders required during early production.

- Reductions in product development cycle times of the order of 40–60 per cent through the concurrent, rather than sequential, design of products and processes.

- Manufacturing costs reduced by as much as 30–40 per cent through having integrated multifunctional teams handling product and process designs.

- Scrap and rework reduced by as much as 75 per cent through product and process design optimization.

Clearly, for any business grappling with improving its agility, concurrent engineering offered obvious answers to the problems associated with accelerating the product innovation cycle. Through concurrent engineering, it was possible not only to cut development lead times, but also ramp-up production more quickly, with some certainty that product manufacturability would not be impaired. And finally, of course, the 'voice of the customer' – if present – should help to provide an assurance that the product that ultimately reached the marketplace was a product that compellingly met customer needs and would find ready buyers.

However, as Winner *et al* also noted – and many others have found subsequently – concurrent engineering was not an instant, easily adopted, panacea. Rather than a destination, it was a journey – and one where success was really only measurable at the end of product development cycles, when products had been taken through from conception and visualization to a successful market launch.

Significant cultural and management changes usually underpinned the successful implementation of concurrent engineering, for instance. And concurrent engineering wasn't a quick fix: a lengthy interval – two to four years, at least – was often needed before benefits were realized from the adoption of concurrent engineering. What's more, concurrent engineering required top-down leadership and involvement in order to succeed, reinforced by training, sponsorship from senior stakeholders and dialogue with relevant parties right across the organization and beyond.

To some extent, of course, that is true of any major organizational change. But unlike, say, approaches such as Six Sigma, concurrent engineering is a *strategy*. It is a *design strategy*, a *product development strategy*, a *manufacturing strategy* and a *supply chain strategy*. And ultimately, of course, an overarching business strategy. Concurrent engineering is not to be undertaken lightly, or without appropriate thought, preparation and investment – at least, not if businesses are to have any success with its adoption.

The evolution of concurrent engineering

To some extent, the adoption of Japanese-inspired manufacturing techniques such as just-in-time and lean production in the 1980s and 1990s helped to popularize the notion of concurrent engineering: fairly obviously, any focus on eliminating waste will soon home in on wastes associated with a lack of communication between supplier and customer, and inadequate dovetailing of designs, business processes and planning cycles (Johnsen, 2009).

More than eliminating waste, however, close supplier involvement in the design process – as we have explored in outline in Chapters 1 and 2, for instance – actually contributes compelling competitive advantage. An influential 1991 study by Kim Clark and Takahiro

Fujimoto of Harvard Business School, for instance, highlighted the role played by early supplier involvement as a key explanatory factor in unpicking the significant performance gap that existed between Japanese and Western automotive manufacturers (Clark and Fujimoto, 1991).

As Bidault, Despre and Butler tellingly point out, the conclusion reached by Clark and Fujimoto's research was that the proportion of a supplier's involvement in the total engineering effort devoted to a new car was 30 per cent in Japan, compared to 16 per cent in Europe and just 7 per cent in North America (Bidault *et al*, 1998). The international comparative study of which Clark and Fujimoto's work was a part was also to play a pivotal role in bringing the advantages of the Japanese approach to a much broader audience when picked up by Womack, Jones and Roos in their famous exposition of lean manufacturing, *The Machine that Changed the World* (Womack *et al*, 1990).

Yet pioneering businesses were experimenting with the concept even before then: as early just-in-time advocate Richard Lubben points out, US manufacturer Xerox launched an Early Supplier Involvement programme as far back as 1977. With the avowed intention of 'producing the best product possible, by bringing together the production experience of the supplier and the design requirements of Xerox's engineering function' (Lubben, 1988), the result was a demonstrable improvement in cost, quality and scheduling. In addition, Lubben pointed out, the resulting enhanced cooperation and close working relationships between Xerox and its suppliers helped to improve design standardization and foster value engineering initiatives.

And yet, even once concurrent engineering began to find favour among businesses, the involvement of suppliers and customers was not automatic, and relatively rare – something that makes the story of the development of Boeing's 777 so interesting. Initially, as MIT's Charles Fine – one of the leading academics and experts studying concurrent engineering – reports, early adoptions of concurrent engineering involved carrying out design and design-for-manufacture operations in parallel: thinking about the design, in other words, but also thinking about the manufacturing considerations that had to be taken into account in respect of that design (Fine *et al*, 2005).

This in turn sparked the development of conceptual tools such as virtual prototyping and rapid prototyping, together with technologies designed to give such ideas concrete form – computer simulation techniques, 3D printing and modelling.

2D and 3D concurrent engineering

Such concurrent engineering approaches have been labelled *2D concurrent engineering*, as opposed to *3D concurrent engineering*, which takes into account supply chain issues in the form of customer and supplier input. This is not to downplay the power of 2D concurrent engineering, or the benefits that it may deliver to organizations. Indeed, even while concurrent engineering was in its relative infancy in the late 1990s, studies were ascribing significant benefits. Ganapathy and Goh, for instance, reported improvements on the order of a 30–60 per cent reduction in time-to-market, a 15–50 per cent reduction in life cycle costs and a 55–95 per cent reduction in engineering change requests (Ganapathy and Goh, 1997).

Empirical investigations into the benefits of 3D concurrent engineering have been far fewer in number, not least because instances of 3D concurrent engineering are themselves fewer in number, which is one of the reasons for the publication of this book, of course, as 3D concurrent engineering is a central tenet of the '4C' model.

Moreover, as Charles Fine again points out, estimating the benefits of 3D concurrent engineering as opposed to 2D concurrent engineering is more conceptually challenging, as those benefits extend not just across the organization adopting such 3D engineering techniques, but also across its supply chain (Fine *et al*, 2005). Think, for instance, of the wide range of benefits that we have already seen stemming from greater integration between a business's design functions and supply chain functions – benefits quite apart from a reduced time to market, life cycle cost reduction and reduced post-launch engineering changes. And when those benefits – reduced waste or reduced risk, for instance – accrue to both supplier and customer, price measurement is obviously going to be problematic.

That said, the distinction between 2D concurrent engineering and 3D concurrent engineering is not helpful on every occasion. In particular, it becomes blurred at the point of supply chain interfaces between trading partners. When an automotive vehicle manufacturer models and simulates the lineside delivery of components from components suppliers, for instance, is that 2D concurrent engineering or 3D concurrent engineering?

Similarly, given the growing use of returnable 'closed loop' special-purpose containers within supply chains, there is obvious merit in modelling and simulating the flows of these containers – partly to avoid having insufficient on hand to meet demand at any one point in the loop, and partly to avoid what is often a more significant problem, namely over-ordering such containers, leading to unnecessarily high capital expenditures and poor utilization. In neither case is this what might be considered 'pure' 2D concurrent engineering, but neither is it pure 3D concurrent engineering.

Irrespective of nomenclature or taxonomy, though, the potential benefits are undoubtedly significant. Wheatley, for instance, points to a simulation study on behalf of one US automotive manufacturer which found that of their US $1.3 million investment in 1,100 such containers, equating to a cost of US $1,200 each, just 47 per cent had moved over the previous 30 days and just 59 per cent over the previous 60 days (Wheatley, 2013).

In a similar example, highlighted by Wheatley, at Nissan's Sunderland car plant in the UK – the country's single biggest vehicle assembly site, producing one in three UK-built cars – the introduction and ramping-up of production of the company's 'crossover' Qashqai sport utility vehicle in the mid-2000s was accompanied by extensive simulation of the links between the plant's lean supply chain and the equally-lean assembly process it feeds (Wheatley, 2008). One focus: optimizing the use of purpose-designed stillages and containers used in materials handling processes, in order to balance supply and demand minimum inventory levels of containers, while taking care to avoid the punitive costs of miscalculation: should the demand for stillages and containers exceed supply at any given time, ordinary wire baskets could be used. But parts can't actually be transported in wire baskets, merely stored in them, so their use necessitates a costly transfer operation, involving both operatives and forklift trucks.

And in yet another example, Wheatley relates how when management at Honda's Swindon car plant was planning the installation of a new engine block machining line in July 2007, the company made a conscious decision to depart from previous practice by using simulation to model the interaction of production and supply chain processes, and simulate their operation under various conditions (Wheatley, 2009).

According to Honda engineer Lee Beggan, 'there was a lot going on in terms of new equipment, conveyors and overhead gantry robots,' (Wheatley, 2009) and Nissan wanted to avoid relying on equipment suppliers' estimates in order to determine if its proposed layout and concept design was realistic in terms of the planned daily output rate, cycle times and buffer sizes. A particular concern: the number of overhead gantry robots required in order to move engine blocks between machining processes, while avoiding either bottlenecks or unproductive stoppages. The conclusion: only one overhead gantry robot was required in order to keep the machining process running, and not two, as had been suggested. The saving was substantial.

Concurrency's strategic advantage

Once again – to stress the point – the precise taxonomy of concurrent engineering entailed in such examples is unimportant. In contrast to the traditional sequentially based approach to turning product designs into finished products, *any* form of concurrent engineering should deliver an improvement. 2D concurrent engineering is an improvement on the traditional approach; 3D concurrent engineering – incorporating supply chain-inspired considerations into the design itself – is a further improvement, as is what we might whimsically call '2½D' concurrent engineering, as seen at Nissan and Honda.

In each instance, the point is to inform the design, development and manufacturing engineering processes by reaching out in parallel, concurrently, to information sources beyond the narrow confines of the design and the design functions themselves.

Sometimes, those information sources will yield facts, insights and capabilities from data stored in people's heads, or suppliers' better or more detailed understanding of their own product technologies, or simply smoother processes resulting from improved communication and collaboration. And sometimes, as we have seen, those information sources can be the results of simulations, experiments and market reactions to prototypes or mock-ups.

In each case, the advantage derived comes not simply from the information per se, but from the fact that the information is available in a timescale that permits the product design itself to be changed in order to take advantage of it. Put another way, concurrency enhances agility by permitting businesses to *increase the number of design iterations that they perform*, with each design iteration taking those businesses another step closer to a state of perfection. Moreover, concurrency also lowers the cost of those designs: as at Boeing with the 777, if a group of designers, customers, manufacturing engineers and supply chain specialists discuss half a dozen design options around a whiteboard in a meeting room, each of those options counts as a 'design' – even if only the surviving selected option is the one design officially enshrined with a design number and a formal drawing.

At the present time, the huge advances in computing power that have been made available through massively parallel processing and cloud technology are enabling manufacturers and designers to undertake iterative design analyses far beyond what was possible for an ordinary industrial company to undertake even a handful of years ago. From computational fluid dynamics to finite element analysis to full-blown virtual reality, the ability to harness high-powered computers or even (literally) an entire supercomputer cluster, is now routinely available at a price that ordinary businesses can afford.

How can a component's performance be improved? How can it be made lighter? Stronger? Thinner? Is it possible to substitute a cheaper material for a more expensive one? Fairly obviously, the more that it becomes possible to undertake multiple 'what if' scenarios in order to answer such questions, the more it becomes possible to produce better, lighter, stronger, less expensive products and components, and accelerate the pace of innovation.

So too, of course, with rapid prototyping, where advances in computer power and material deposition technologies now enable companies to produce either virtual 3D prototypes for designers and prospective customers to explore, or physical prototypes, produced at a fraction of the cost of conventional modelling techniques, and in a fraction of the timescale. Using techniques such as stereolithography (SLA), selective laser sintering (SLS) and fused deposition modelling (FDM) – the latter being more colloquially known as 3D printing – prototype parts and components can be produced directly from a designer's CAD files, using equipment that costs a mere fraction of prices prevailing just a few years ago. Rapid prototyping machines that might have cost £200,000 or so 10 years ago are now available at a price one-tenth of that.

That said, the advances that have taken place in rapid prototyping in recent years are not without a certain irony: just as rapid – and inexpensive – prototyping becomes an attainable and affordable reality, the growth in the use of simulation is driving down the need for such physical prototypes.

Agility through postponement

Concurrent engineering isn't the only means open to a manufacturer wanting to improve its competitive position through enhanced agility. An alternative – and even complementary – strategy is to make use of *postponement* instead.

Postponement isn't difficult to understand. Essentially, it relies on designing or manufacturing a product so that *as much* **customer configuration** *as possible can take place as late as possible*. Suppose, for example, that a business sells a product that involves a lengthy component manufacturing process with long lead times – and yet most variants of the product contain a core subset of components that are largely the same. Under such circumstances, it makes sound competitive sense to try to offer the customer a faster order completion time than would otherwise be possible by holding stocks of completed components and assembling the finished product to order, once the precise details of the customer's order are known.

Or, to consider another example, imagine a situation where completed products are largely the same from customer to customer, but that selectable or customer-specific options apply in the case of associated items that are shipped with the product, or which are required in order to use it. Again, it makes sense to hold stocks of the completed product but apply the customer-specific or otherwise selectable options as late as possible, once order details are known.

This was the basis of the country localization that we saw in Chapter 2, of course, in respect of Hewlett-Packard's global desktop printer business: printers were held in inventory as completed units, with country-specific items such as power cords, product manuals and labelling added subsequently, when it was clearer which printers were being sent for sale in which markets. Through a strategy of postponement, Hewlett-Packard was able to strip out complexity, reduce inventory, improve the inventory availability of country-specific SKUs, improve customer service levels and reduce lead times.

Postponement in practice

Computer manufacturer Dell is another business that has become recognized for leveraging postponement adroitly within its supply chains. More than improving operational effectiveness, though, Dell has used postponement to create innovative business models and customer propositions, cleverly blending together the power of its build-to-order business model with its ability to exploit its purchasing power and economies of scale to procure industry-standard items such as keyboards and monitors.

In the early 1990s, for instance, as Dell was rapidly expanding its global footprint, it established a 1,900-employee manufacturing plant in the Irish city of Limerick, intended to build built-to-order desktop computers, laptop computers and servers for the European, Middle East and African markets (Wheatley, 1999). The location – a small country on Europe's periphery – wasn't the most logical choice from a logistics network planning point of view but was instead influenced by tax breaks and grants from the Irish government and its Irish Development Agency. Postponement, though, was the trick that made the location work.

Dell's subsidiaries in each of the national markets in which the company operated would take orders from customers and then transmit them to the Limerick factory for manufacture. Within 24 – or occasionally, 48 – hours, the machines in question would be progressing down the production lines in Limerick. But only the machines themselves: monitors, keyboards, loudspeakers and so on would be shipped directly from suppliers to storage locations as close as possible to the customer, where they would be held on a consignment stock basis, awaiting call-off by Dell to meet particular customers' orders, and only united with the Limerick-manufactured computer when sent for final delivery to the customer.

For computers destined for the UK market, for instance, such items were held in Liverpool. But while the physical Liverpool stockholding generally amounted to around 20 days' worth of such items – monitors in particular were seen as subject to supply chain irregularities – the monitors and speakers associated with each computer order only went into Dell's inventory for around half a day. After this, they would either go into accounts receivable (in the case of computers for the business market) or cash, in the case of consumer computers paid for by credit card at point of despatch. Given that suppliers were paid around 45 days after their products were taken into Dell's inventory, this meant that Dell found itself in the enviable position of paying suppliers some six weeks after it had itself been paid.

The world's automotive industry is also a major adopter of postponement strategies, relying on them to maximize end-customer configurability while minimizing finished vehicle inventories – a particular challenge when vehicle manufacturing and demand forecasting processes might take place on the other side of the world from markets such as North America and Europe. Consequently, many have taken a strategic decision to tailor vehicles for particular national markets post-manufacture, rather than during the manufacturing process itself. As a result, they can manufacture simpler, more standard models and take the decision about customizing them later, when more precise estimates of the demand for particular options is available.

Mazda North America, for instance, has a policy of customizing for the US market at the port of entry, a process that the company calls 'port accessorization'. The logic: most of the vehicles for Mazda's world markets are manufactured in Japan, but not all markets want or need the same product features. A practice of having port-installed accessories or customer-selectable options allows the company to address individual market needs without impacting the manufacturing plant's assembly processes. Moreover, it frees it from the burden of having to forecast the demand for particular options: it need only forecast the overall demand for generic models, knowing that customizing to individual dealer requirements can take place much closer to the point of delivery.

When dealers order vehicles, they have the option to add accessories such as remote starting, spoilers, satellite radio and roof racks. The completed vehicles will be manufactured and shipped without these options, which will then be fitted once the vehicles reach their port of destination, usually by specialist compound management firms. Installation time typically averages around half an hour of accessory installation time per vehicle, and often has the advantage of a cheaper labour rate than that obtainable at the dealer. Only if the global penetration rate for a particular product feature is high enough will installation then be driven back into the assembly plant – but otherwise the emphasis is on keeping things at the factory as simple as possible.

Design for postponement

As with concurrent engineering, though, postponement is a strategy, not an instant panacea. A business cannot suddenly decide to switch to using postponement overnight; instead, postponement may require significant changes to product designs, procurement manufacturing processes, sales processes, logistics and labelling processes, and overarching business processes. Granted, not all of these may require changes: sales processes, for instance, may be unaffected, with customers simply unaware that late-stage configuration lies behind a manufacturer's ability to offer short lead times on products that may take weeks to manufacture. But in general, it is worth considering the possibility of making changes in all of these areas, in order to exploit to the maximum business benefits that a new-found flexibility confers.

Most obviously, though, all of these changes start with product design: if a product has not been designed with postponement in mind, it can sometimes be difficult – if not almost impossible – to adopt a postponement strategy. To choose another example, electrical appliances routinely exploit postponement in the form of the power leads with which they are supplied: twin-pin or three-pin flat pin Type A or Type B in North America and Japan, for instance; two-pin or three-pin round pin Type C, Type E or Type F plugs for Europe; three-pin oblong pin Type G plugs for use in the United Kingdom, Ireland, Malaysia and Singapore, and three-pin angled flat pin plugs for use in Australasia.

From a physical postponement point of view, the correct plug can be supplied at almost any point after product manufacture – at the packing stage, for instance, or even in a global regional distribution centre. But the supply of the correct plug is only part of the story: the product itself must be capable of switching between the two main voltages used in consumers' homes and offices: 120 volts in North America and Japan, and 220–240 volts everywhere else.

Next, manufacturers will want to make sure that manufacturing and procurement processes are correctly configured to suit their postponement strategy. To choose another example, consider Zara's innovative use of undyed fabric, which provides them with the flexibility to decide the precise as-sold colour at a point much closer to the point of sale, when it has a much firmer view of each season's fashionable colours and consumer preferences.

If an item of clothing has been designed from the outset to be, say, red, with red fabric duly procured for its manufacture, then it will be impossible to subsequently re-dye this fabric to a lighter colour such as white or yellow. So buyers need to know that they will be buying undyed fabric, and manufacturing processes need to reflect the inclusion of a dyeing and finishing process much further along the production process than had hitherto been the case.

As an aside, note that the added flexibility offered by postponement does not necessarily have to come at an additional cost in such instances. While the dyeing and finishing of fabric may be an additional operation for a manufacturer to take on or coordinate, procurement economies are available through buying a greater quantity of undyed fabric, as opposed to fragmented purchases of individual colours; dyeing and finishing would happen anyway – all that has happened is that the point in the production process at which it takes place has been shifted; and it can be reasonably assumed that the ability to more precisely gauge which colours are selling well (and which aren't) will result in lower levels of discounting and obsolescence, not to mention higher levels of availability for items in colours that are selling well.

So too with logistics and labelling processes. It is easy to see the advantages of postponement taking place as close as possible to customer demand, in regional distribution centres and warehouses.

But time must be allowed for the postponement processes within the overall lead time; operatives trained to perform them correctly; suitable space found in which they can take place; measures taken to prevent theft and losses; systems put in place to replicate the quality assurance practices attainable in a factory environment; and supply chains reconfigured so as to send the appropriate parts to, say, globally distributed regional distribution centres and warehouses. In the case of power leads for appliances, for instance, supply chains that currently terminate at, say, a factory in China or Japan might instead have to ship the power leads in question to a warehouse on an industrial park adjacent to Schiphol airport (Amsterdam). None of these are insurmountable challenges, but if experience teaches anything, it is that warehouses do not easily lend themselves to manufacturing operations.

In summary then, as with concurrent engineering, postponement is not a quick fix. It is a powerful strategy, but not a rapid one in terms of delivering benefits. And, once again, its proper consideration must first take place on the drawing board.

CASE STUDY   Johnstons of Elgin: agility as a counter to cost competition

A manufacturer of upmarket classically styled cashmere and lambswool clothing garments, Johnstons of Elgin can trace its history back to 1797, when Alexander Johnston first took a lease on a woollen factory at Newmill in Aberdeenshire, Scotland. Over two hundred years later, the mill of Johnstons at Elgin is still on the same site and is the UK's last remaining vertically integrated woollen mill – the only woollen mill in the country to still carry out all the manufacturing processes from the blending, carding and spinning of wool to the knitting and weaving of finished fabrics and garments, all at a single location.

Originally a producer of 'estate tweeds' – a derivative of tartan – the company gradually branched out into the woven cashmere business, importing cashmere and developing a range of fine woven cashmere cloths. Much later, in 1973, Johnstons entered the cashmere knitting business, with a separate cashmere knitting factory located at Hawick in the Scottish borders.

For many years cashmere-based products had tended to be highly priced and as a result were purchased only by more affluent consumers. But with the onset of globalization, new sources of low-cost competition began to emerge in the 1990s. Products labelled as 'cashmere' could be sold in supermarkets in western countries at a fraction of the price that traditional manufacturers and retailers were charging. In 2008, for instance, a cashmere pashmina could be bought in retailer Tesco for £29, compared to £200 or so in a store such as Harvey Nichols. Admittedly, many of these low-cost imports were not of the same quality, and often contained only just enough cashmere wool to enable them legally to be labelled as cashmere – but despite this, they quickly had a severe impact on the sales of UK-produced cashmere products.

Many traditional manufacturers were not able to withstand this competition and left the industry. Johnstons itself was not immune from these competitive pressures, and in 2006 it saw its profits fall from £2.2m in the previous year to £336,000.

As part of the fight back, the company decided to exploit the fact that design had become a much more important aspect of its business. Originally largely a menswear business with highly stable products with long life cycles, such as fabrics for suit making, Johnstons' product range had shifted over time, making it now very much a womenswear business, with a higher fashion content and with much shorter life cycles.

At the same time, there had been a transition from a business producing mainly standard products on a repetitive basis, to a much more customized product base, producing lines for the company's own retail outlets, as well as 'own label' product ranges for major fashion houses such as Hermès.

Accordingly, Johnstons' management recognized that a strategy of becoming a design-led company could provide it with a powerful platform for competing against low-cost country sources, using design as a differentiator. At the same time, though, it recognized that just as design had become of increasing importance to the overall business, it had also become a much more critical element in the product development process – and also that, as is common in the textile and apparel industry, the time that it took to go from initial design to finished product in the market was often very lengthy.

Partly, the length of the product development process had been caused by the inflexibility of the traditional production and finishing processes. This had been significantly eased, though, by a number of changes and innovations that had occurred in Johnstons' manufacturing processes, such as the introduction of 'hank and package' dyeing of yarn, and the purchase of new equipment that could produce and process fabric in smaller batches.

But the design process had seen fewer changes and innovations, and there was an emerging view that it needed to radically change if it were not to hold back the business as it fought back against low-cost imports. It was not sufficient to be innovative in design if new products could not be introduced rapidly and production adjusted quickly to match uncertain demand.

One issue: the delay caused by the need to produce samples of the finished fabric for clients, and then often make frequent changes to the design of the product at the request of those clients. Not only did these delays add significantly to the cost – the cost of a sample might be in the region of £80 a metre – but also it meant that the time to market was extended.

Another problem was the fixed design cycle that Johnstons' design process followed. For its own range of products (as distinct from those manufactured for other customers) the business followed a regular cycle: initial work on new designs and colour ideas began in February, with June as the deadline for the first review of new product ideas and a final 'sign off' at the end of August. Products would then appear in the shops the following April/May.

But for those products that Johnstons manufactured for other customers, such as fashion houses or retailers, the competitive pressures meant that the design cycle had to be shorter and more flexible. These customers, who were of growing importance to Johnstons, were highly demanding in their requirements and often made late changes to product designs and specification. In addition, retail customers such as Burberry had increased the number of seasons for their range changes, generally from two a year to four. They were also requiring the introduction of new colours in mid-season, again with a requirement for pre-production samples.

The most immediate challenge in accomplishing this was the availability of capacity: whereas in the past the focus had been on reducing capacity to take costs out of the business, there was now a requirement to either find better ways to use existing capacity, or possibly to access capacity elsewhere. Either way, the shift in business strategy signalled the need for agility, rather than a focus purely on cost and a lean operating model. Nor could the additional capacity needed to create this agility simply be purchased as capital equipment investments: the problem was not so much the number of machine hours available, but rather the availability of skilled people. As the workforce was gradually ageing, the pool of experienced workers was diminishing, which was particularly the case with those production tasks that involved hand sewing.

One way that Johnstons sought to improve their responsiveness to demand was by making greater use of outsourcing as a strategy, consciously separating the more predictable product lines in the range from the less predictable parts of their business. Consequently, a partnership was established with a mill in Mauritius, which was used to produce a number of Johnstons' higher-volume lines. This approach offered the twin merits of combining the advantages of low-cost manufacturing with an associated freeing-up of the core Scottish operations for the more fashion-oriented, short life cycle lines. Here again, outsourcing further enhanced Johnstons' agility through a decision to outsource ancillary operations such as finishing to other UK companies.

In addition, to mitigate the spiralling cost of raw cashmere, the company increased options for customers to buy blended products, such as cashmere mixed with the less expensive extra-fine Merino wool. Similarly, as cashmere is a dollar-based commodity, the company's designers collaborated with worldwide sales agents to lessen the impact of exchange rate fluctuations, by consciously designing products with a geographic focus on regions where prevailing exchange rates were likely to make Scottish products more attractive during the fashion season in question. Furthermore, Johnstons began exploring new technology that printed patterns digitally onto scarves manufactured from a generic woven cloth. This enabled it to rapidly introduce new styles while taking advantage of the manufacturing benefits associated with the simplified dyeing, weaving and finishing of a generic cloth.

Checklist
Questions for design professionals

☐ How does 'the voice of the customer' currently reach your business's product design function? Are these contacts formal, informal and linked – or not – to the development of particular new products?

☐ Do you know if your business makes use of concurrent engineering? If not, why not? If concurrent engineering *is* in use, is the supply chain involved? And if not, why not?

☐ Do your product design processes formally consider aftermarket servicing and maintenance considerations? If so, how is this data captured? And how is it incorporated in product design decisions?

☐ What opportunities exist for your business to adopt digital simulation techniques? Have these ever been considered? If not, why not?

☐ What opportunities exist for your business to adopt rapid prototyping techniques? Have these ever been considered? If not, why not?

Checklist
Questions for supply chain professionals

☐ Do you know if your business makes use of concurrent engineering? If not, why not? If concurrent engineering *is* in use, is the supply chain involved? And if not, why not?

☐ What business benefits might stem from the adoption of postponement practices? Have these ever been evaluated, or postponement considered? Which stages of your production, labelling or logistics processes might lend themselves to the adoption of postponement?

☐ How does 'the voice of the customer' currently reach your business's product design function? Are these contacts formal, informal and linked – or not – to the development of particular new products?

☐ Are suppliers routinely involved in new product design and development processes? Have they ever been asked to be? And do formal processes exist for identifying which suppliers are well-placed to assist in terms of their competitive edge in product design and product technologies?

☐ Has your business ever undertaken formal modelling of supply chain interactions with suppliers, especially at the point to interfacing with the manufacturing function – '2½D' concurrent engineering, as it is termed in this book? Which specific aspects of your supply chain might benefit from such modelling?

References

Bidault, F, Despres, C and Butler, C (1998) New Product Development and Early Supplier Involvement (ESI), *International Journal of Technology Management*, 15 (1/2), pp 46–69

Christopher, M (1998) *Logistics and Supply Chain Management: Strategies for reducing cost and improving service*, 2nd edn, FT-Prentice Hall, London, p 49

Clark, KB and Fujimoto, T (1991) *Product Development Performance: Strategy, organization, and management in the world auto industry*, Harvard Business School Press, Boston

Fine, Charles H, Golany, B and Naserald, H (2005) Modelling Tradeoffs in Three-Dimensional Concurrent Engineering: A goal programming approach, *Journal of Operations Management*, 23 (3–4), pp 389–403

Ganapathy, B and Goh, CH (1997) A Hierarchical System of Performance Measures for Concurrent Engineering, *Concurrent Engineering: Research and Application*, 5 (2), pp 137–43

Glende, WL (1997) The Boeing 777: A Look Back, NATO STO, Seattle [Online] available at: https://pdfs.semanticscholar.org/0b62/33d8f075be 5cb74af32968b219863ec3d703.pdf [accessed 22 January 2018]

Johnsen, TE (2009) Supplier Involvement in New Product Development and Innovation: Taking stock and looking to the future, *Journal of Purchasing and Supply Management*, 15 (3), pp 187–97

Lee, H (2004) The Three A's of Supply Chain Excellence, *EE Times* [Online] available at: https://www.eetimes.com/document.asp?doc_id=1240026 [accessed 24 January 2018]

Lubben, R (1988) *Just in Time Manufacturing: An aggressive manufacturing strategy*, McGraw-Hill, New York

Wheatley, M (1999) Compaq vs. Dell, *Logistics Europe*, September, p 18

Wheatley, M (2008) Simulation, Better Forecasting, and Advanced Planning Smooth Lean Transition to the Supply Chain, *Manufacturing Business Technology*, January

Wheatley, M (2009) Simulating the factory, *Engineering and Technology*, August

Wheatley, M (2013) Virtual Reality, *The Manufacturer*, 4 September, p 70

Winner, RI, Pennell, JP, Bertrand, HE and Slusarczuk, MM (1988) *The Role of Concurrent Engineering in Weapons System Acquisition*, Institute for Defense Analyses, Alexandria, VA

Womack, J, Jones, D and Roos, D (1990) *The Machine That Changed the World*, Simon & Schuster, London

Further reading

Matthews, P and Syed, N (2004) The Power of Postponement, *Supply Chain Management Review*, 8 (3), pp 28–34

Serling, RJ (1992) *Legend and Legacy: the story of Boeing and its people*, St. Martin's Press, London

Spitz, W, Golaszewski, R, Beradino, F and Johnson, J (2001) *Development Cycle Time Simulation for Civil Aircraft*, NASA/CR-2001-210658, pp 2–7

Wheatley, M (1998) Pile it low, sell it fast, *Management Today*, 1 February, p 62

Product design and sustainability 05

Introduction

Over the past quarter of a century, the sustainability agenda has become increasingly important for businesses. Increasingly, too, recognition has dawned that *sustainability is a supply chain issue*, rather than something to be considered in isolation within the enterprise itself.

This chapter looks at the role that the product design process plays in improving businesses' sustainability performance; at *how product design decisions affect the environment*, together with the people and animals living in it; and at *how designers often need to work with supply chain partners* in order to achieve their sustainability objectives. It explores the *major sustainability impacts that product design decisions can have* and makes the point that *sustainability need not come at the expense of profitability*.

Partnership for good: supplier collaboration in action

Determined to improve its sustainability performance, pharmaceutical giant GlaxoSmithKline decided in 2012 to take steps to reduce its carbon footprint – but quickly reached two uncomfortable conclusions.

First, that 40 per cent of its carbon footprint lay not within its own operations but in those of its supplier base, from whom it spent £2 billion buying materials. And second, that individual suppliers' contributions to that carbon footprint were highly fragmented, with no single supplier making up more than 1 per cent of it.

'The task looked huge: we were going to have to engage with an enormous number of suppliers if we were to make a material impact in helping to reduce their carbon footprint – and so help us to reduce our own overall footprint,' subsequently recalled Matt Wilson, Head of GlaxoSmithKline's global Environmental Sustainability Centre of Excellence (GlaxoSmithKline, 2015).

A consulting firm was duly engaged to talk to a cross-section of suppliers, probing their focus on improving energy efficiency and trying to establish if they were open to the idea of collaborating with GlaxoSmithKline in order to identify and deliver improvement in energy efficiency.

'Trying to do it through our procurement organization would lead to a commercially-oriented conversation, which sets the wrong tone,' explained Wilson. 'By going through consultants, it acted as a buffer, and by providing suppliers with the ability to respond anonymously if they liked, we further contributed to their comfort level' (GlaxoSmithKline, 2015).

Seventy per cent of respondents reported that saving energy was important to them, which was a step in the right direction. But only 35 per cent went on to report that an actual programme was in place that was actually aimed at delivering energy savings. This was much less encouraging and clearly something that was going to have an impact on GlaxoSmithKline's own carbon-reducing aspirations and achievements.

Yet there were also signs for optimism. Around 85 per cent of suppliers had been happy to respond openly, rather than taking advantage of the proffered anonymity. Moreover, some 90 per cent were happy to engage in collaboration programmes aimed at improving energy efficiency. And perhaps best of all, a significant number of suppliers' responses had come not from their commercial arms, but elsewhere in their organizations.

In short, realized GlaxoSmithKline management, if the conversation could be moved from one taking place between the commercial teams of GlaxoSmithKline and its suppliers – and engage directly with people who were much closer to sustainability issues – then there were grounds for hoping that a much richer conversation could take place.

But how? It turned out that a ready-made model for that direct engagement and collaboration already existed, GlaxoSmithKline management realized, in the shape of Walmart-owned supermarket chain Asda's Sustain & Save Exchange (2degrees, nd). This was a private online community of Asda's suppliers and Asda representatives, collectively focused on sustainability and cost-efficiency issues. Launched in 2012, the community aimed to share sustainability best practice among members, with any cost savings firmly ring-fenced from Asda's commercial team.

Consciously facilitated by an independent third-party specialist provider of such online communities, 2degrees, the community primarily took the form of interactions between suppliers, rather than interactions between suppliers and Asda representatives. Suppliers asked questions of other suppliers, and either facilitated or took part in various knowledge-sharing initiatives, such as site visits, workshops, online training sessions, webinars and expert masterclasses. Suppliers might ask online for advice on, say, LED lighting, and other suppliers would respond with their own experiences, offering impromptu site visits or referrals as appropriate.

As the vast majority of interactions were between suppliers in very different businesses, selling very different products to Asda, a high degree of openness existed. And while there was no compulsion to join, the appeal of gaining free access to information on improved sustainability practices – and often, the consequent cost savings – had quickly persuaded some 300 suppliers in 14 countries to sign up, collectively representing some £7 billion in over-the-checkout Asda sales of fresh, chilled, ambient and frozen foodstuffs.

In short, from the adoption of LED lighting to biomass boilers, and improved air compressor efficiency to reduced packaging waste, Asda could point GlaxoSmithKline to a rich seam of examples of shared best practice that would have been difficult to engender in any other way.

GlaxoSmithKline management were quickly convinced, launching their own version of the community – again facilitated by 2degrees – in 2014. A year later, the GlaxoSmithKline Supplier Exchange embraced some 270 of the pharmaceutical giant's suppliers of direct materials, representing some £1 billion of spend, a figure that as of 2017 had grown to 320 suppliers in 45 countries around the world. The stated objective: reducing GlaxoSmithKline's carbon footprint by 25 per cent by 2020 and helping it to build a carbon-neutral value chain by 2050.

The supply chain challenge

Once, of course, such collaboration and cooperation would have been unheard of. Suppliers, freely collaborating and working with... *other suppliers*? Businesses working with suppliers on sustainability issues at all would also have been unthinkable.

But the world changes. And just as human actions have indisputably harmed the environment over the past two or three centuries, human action in the form of supply chain collaboration is now attempting to redress some of that damage.

The work being done by GlaxoSmithKline and Asda shines a spotlight on one of the most difficult aspects of conventional 'longitudinal' supply chain collaboration: however open the relationships, however willing the collaboration, the mere act of communicating and collaborating does not of itself guarantee to provide answers or results. Put another way, a customer and a supplier, working in isolation, do not possess all the answers.

But other customers and suppliers, elsewhere, *might* possess those answers. And as product designers and supply chain managers become ever more demanding in their search for improvements in sustainability performance, an ability to access the insights, expertise and experience of those other customers and suppliers becomes ever more important.

On their own, too, product designers' insights into improved sustainability are also constrained by their own knowledge, expertise and experience. They may want to design a more environmentally friendly product, but recognize that this will require skills and technologies not present within their own orbit. That is where the supply chain, and their supply chain colleagues, come in. And in this chapter, we explore some of the sustainability issues that those conversations might embrace.

Design decisions matter

The central tenet of this chapter is that product design can have a major impact on the environment and on human civilization's sustainability performance. Not as a 'one off' special project, or as an optional extra to be selected at will by designers and their customers, or as a piece of token greenery to be pointed at in a company's sustainability report – but as a conscious aspect of everyday design activity, a mental agenda underpinning the entire product development process.

As Clare Brass, Flora Bowden and John Moseley put it in a pivotal paper published by the UK's Design Council (Brass *et al*, 2007):

Design for sustainable [development] is neither an add-on, nor an elite area of design. Design for sustainable development is the process by which all designers can improve the social, economic and environmental impact of their work.

That this is desirable should come as no surprise, and while it is not the intention of this book – or this chapter – to promote a green agenda for the sake of it, it is difficult to ignore the growing body of evidence that human decisions – and in particular, product design-related human decisions – are having an adverse effect on the world that we live in.

And the evidence continues to mount. Back in 1962, when influential writer Rachel Carson published *Silent Spring*, the idea that synthetic pesticides such as dichlorodiphenyltrichloroethane – better known as DDT – was killing wildlife was fiercely resisted by vested interests such as chemical companies (Carson, 1962). Nevertheless, Carson's influential work ultimately led to the creation of the United States' Environmental Protection Agency and resulted, in 1972, in a ban on DDT's use on agricultural land. A similar ban in other countries soon followed.

1972 was a formative year in other ways, too. Funded by the Volkswagen Foundation and commissioned by the Club of Rome, Meadows *et al*'s seminal *The Limits to Growth* (Meadows *et al*, 1972) raised important concerns about resource consumption and the depletion of the world's finite resources – oil, gas, metals, fresh water and so on. Based on complex computer simulations and modelling, it served as a powerful wake up call, albeit a controversial one. Forty years on, the model's central projections are reported to be still in line with real-world data – not least 'peak oil', which although deferred, increasingly seems inevitable as global extraction exceeds the discovery of economically viable accessible new oil fields.

Twenty years after *The Limits to Growth* came the 1992 United Nations Earth Summit, held in Rio de Janeiro. For the first time, it put sustainable development on a multinational, cross-border footing. One hundred and seventy-two governments attended, with 116 of these sending their heads of state or governments. Critically, it lay the footings for what became the United Nations Convention on Biological Diversity and the United Nations Framework Convention on Climate Change, both of which were formally opened for signature at the summit.

Turn the clock forward to today, and the extent of the progress that has been made is very evident, albeit unevenly across the world. Yet against the undoubted progress that has been made in Western Europe and North America must be set the generally backwards moves in countries such as China. For while China's politicians have made rather more environmentally friendly noises of late, with a sharper focus on water quality, air quality and a clampdown on pollution, significant amounts of environmental damage have been done in the drive for rapid industrialization and export growth.

The Economist, for instance, pointed in 2017 to a Chinese government national soil survey, which showed that 16.1 per cent of all soil and 19.4 per cent of farmland was contaminated by organic and inorganic chemical pollutants and by metals such as lead, cadmium and arsenic – with the latter two found in 40 per cent of farmland (*The Economist*, 2017). This amounts to 250,000 square kilometres, it adds, an area equivalent to the entire area of arable farmland in Mexico. Thirty-five thousand square kilometres were so polluted that the official view was that no agriculture should be allowed on it at all. Cadmium is released during the smelting of iron, copper and lead, with the outpourings from China's ubiquitous smoky and inefficient smelters and blast furnaces being well-known to visitors to the country – despite the short-lived improvements in air quality achieved in the run-up to the 2008 Beijing Olympics as such plants were closed down or moved.

And it is precisely such disparities in environmental performance, say sustainability campaign groups, that point up the importance of tackling environmental harm at source – not the factory or manufacturing process, but at the design stage. For just as 80 per cent of a product's cost is 'baked in' during product design, so too are 80 per cent of products' environmental impacts, point out Brass, Bowden and Moseley, being reflected in the choices made regarding products' material composition, manufacturing process, energy efficiency and packaging (Brass *et al*, 2007).

Put another way, an indeterminate amount of the improvement seen in the environment in Western Europe and North America over the past two decades has not been achieved by *eliminating* harmful environmental effects, but simply by *relocating them* thanks to globalization and the outsourcing of manufacturing to factories in low-cost economies. These low-cost economies may be exporting their production to us in the West, but we are surely exporting our environmental damage to them.

Moreover, this *environmental* damage caused by outsourcing manufacture to low-cost economies can also be mirrored by *societal* damage: harm not to the environment as such, but to the people living within it, or dependent upon it.

Sometimes, such harm attracts unwelcome headlines and reputational damage. Ask retailers such as Walmart, Benetton, Matalan and Primark, slammed by consumer activists after using garment factories housed in the unsafely built nine-storey Rana Plaza building outside Dhaka, Bangladesh, which collapsed in April 2013, killing over 1,100 workers. Or ask computer giant Apple, still smarting after a spate of high-profile suicides at its Foxconn manufacturing contractor in 2010 – 14 employees died after throwing themselves off rooftops – or sportswear company Nike, still bearing scars some 15 years after some of its shoes were found to be produced by children working in sweatshops in the developing economies.

But sometimes, the harm doesn't attract such headlines, despite the societal damage – and individual harm – being even greater. In parts of India and Bangladesh, for instance, the dyeing of textile fabrics (often for Western clothing and supermarket chains) has caused extensive water pollution through unregulated dumping of waste water. Despite the use of non-polluting or low-polluting dyeing technologies (such as using recycled water, using an alternative carbon dioxide-based dyeing process, using biodegradable 'natural' dyes and designing products based around polyester fabrics, rather than cotton), the practice continues.

As Britain's *Guardian* newspaper observes, in Tirupur, India – home to scores of factories and workshops where workers dye materials for t-shirts and other garments marketed around the world – local dye houses have long dumped waste water into the local river, rendering groundwater undrinkable and local farmland ruined. 'Despite tougher regulations, a watchful local press, and the closure of companies in non-compliance, water pollution has festered. The city's 350,000 residents, not multinational textile companies, pay the price,' it points out (Kaye, 2013).

More recently, attention has turned not to the dyeing of fabrics, but the choice of fabrics themselves. Viscose, also known as rayon – and often seen as a cheaper alternative to both cotton and polyester – is easily portrayed as a 'natural' fabric, being produced from cellulose or wood pulp, usually (although sadly not always) sourced from specially grown fast-growing trees, as with board and paper production. The harm to the environment, and to society, comes not from the raw material, though, but from the industrial process required in order to turn that raw material into the end product.

Among other chemicals involved are sodium hydroxide (caustic soda), sulphuric acid and carbon disulphide – the latter a known toxin that has been linked to coronary heart disease, birth defects, skin conditions and cancer. Widespread release of these into the local environment surrounding manufacturing plants in certain low-cost economies has produced high levels of water pollution, rendering drinking wells unusable, the Changing Markets Foundation has alleged, again citing well-known Western fashion retailers as buyers of the viscose in question (Hoskins, 2017).

The bottom line: while the principal focus of this chapter is on sustainability in the context of the environment, it is sometimes difficult to disentangle environmental sustainability from societal sustainability. From a purely design perspective, it is best to be pragmatic and acknowledge the limits of what design decisions, in conjunction with the supply chain, can affect.

Issues such as, say, child labour or unsafe or otherwise exploitative working conditions are to be deplored, but there is little that a designer can do to impact them. Instead, these are more properly the remit of the procurement function, in terms of sourcing and allocation decisions. But issues such as, say, the choice of fabric dye, or fabric dyeing process, or viscose versus cotton – here, the product designer can make decisions that *can* make a difference.

Difficult choices

Not all such decisions are as straightforward, though. Consider, for instance, the multiple impacts of obsolescence on the environment. Almost unarguably, the best way to minimize the impact of manufacturing industry on the environment is to reduce the *demand* on manufacturing industry – together with its resource consumption, industrial processes and waste by-products – by making products that last longer. With products that last longer, and which therefore need replacing less frequently, the slowdown in manufacturing activity then ripples backwards right through the supply chain – less demand, so less production, so less resource requirement, less transport and so on.

Here again, though, there are societal impacts. First, obsolescence is not a simple 'yes or no' decision. Contrary to anti-consumerist belief, designers do not (in most businesses, anyway) sit at their desks deliberately designing products that will wear out or fail sooner. But they *do* design products to 'price points', creating product and material specifications that allow products to be manufactured and sold at a given target price that will deliver a given level of profit.

And they do this for a reason: consumers have varying degrees of wealth, and wildly varying incomes and budgets, and a sensible manufacturer (with the obvious exception of luxury brands and niche marques) does not willingly artificially constrain itself by deliberately excluding potential markets. Making products last longer might therefore be expected to add cost to products and therefore price, taking them outside the reach of certain market segments.

Likewise, as economists will confirm, there is another equally obvious impact of longer product lives – namely a less active manufacturing industry and a smaller workforce, as fewer products need to be manufactured to serve the needs of a given set of consumers over a given period. Obsolescence might be bad for the planet, in other words, but it is perversely *good* for employment, incomes and gross domestic product.

Moreover, reducing obsolescence is not always a practical or desirable proposition: 'fitness for purpose' comes into the equation. With *consumer durables*, product longevity makes sense and is generally desirable, subject to affordability and a feature set that is relevant over the long term. Had a 1960s-era washing machine lasted into the present day, for instance, it might still be functional in a 1960s sense, but would be unlikely to satisfy the modern consumer, offering what would now be regarded as a fairly rudimentary feature set. Similarly, with cars and many other durables: people may delight in showing off their vintage cars at enthusiasts' rallies, but few would these days elect to use such vehicles for everyday driving.

Food products are another case in point. Apart from an obvious interest in reducing avoidable food waste, it is not necessarily sensible for foodstuffs to be made to last longer: a can of beer will be drunk when the consumer decides that they need a drink, and it can obviously only be drunk once. Likewise, with any number of other consumer consumables, from cleaning products to stationery and utensils: their purpose is to be consumed – and then, when consumed, discarded.

Yet even so, as product designers wrestle with some of product longevity's less obvious ramifications, there is a further card up their collective sleeves. For even once a product has been used and discarded, or simply rendered irrelevant for modern-day needs, it makes sense to design it in such a way that it can be readily recycled or reused in another form. So while the contents of a can of beer may have been consumed, the can lives on in a collecting bin, to be melted down and reformed once more as a can. So too with the plastic drinks bottle, the plastic containers used for foodstuff and cleaning products, and much everyday paper, plastic, glass, card and metals – the very categories that most domestic recycling schemes focus on.

The circular economy

Yet it is possible to go much, much further than this. A concept known as the 'circular economy', and generally first formally associated with environmental economists David Pearce and Kerry Turner, argues that today's *linear* economy – with supply chains stretching from raw materials to finished goods – should instead be replaced by a more sustainable *circular* economy, where the raw materials from which finished goods are produced are largely (or even exclusively) made up of recycled materials from previous generations of those same (or indeed, other) finished goods (Pearce and Turner, 1989).

Today popularly associated with the work of the Ellen MacArthur Foundation and a number of other environmental activist organizations, advocates for the circular economy point out the cost, resilience and sustainability benefits of products that from the outset have been designed for such circularity.

By designing products made entirely or largely from recycled paper, cardboard, plastic, aluminium or steel, for instance, product designers are utilizing a waste stream of materials that would otherwise have to be disposed of. And while the cost of such waste is not 'free' – it must be collected, consolidated and processed – there should clearly be economies when compared to the cost of extracting and processing virgin raw materials on the other side of the planet.

Moreover, by disrupting (and minimizing) the stream of incoming virgin raw materials from which their products would otherwise be manufactured, manufacturers are not only delivering on their cost and sustainability agendas, but are also helping to make their supply chains more resilient, by tapping into a 'near shore' supply chain of recycled material that is in theory both less subject to price volatility *and* less subject to natural disasters, geopolitical impacts or long-distance freight disruptions.

A report by the Ellen MacArthur Foundation, *Towards a Circular Economy: The Business Rationale for an Accelerated Transition*, estimates that by adopting circular economy principles, Europe could create a net benefit of €1.8 trillion by 2030 (Ellen MacArthur Foundation, 2015). The cost of recycling and reusing parts of mobile phones could be reduced by 50 per cent per device, for example, if only the mobile phone industry made phones easier to take apart and recycle, and offered consumers incentives to return their old mobile phones rather than discard them. Likewise, the packaging, processing and distribution costs of bottled and canned beer could be reduced by 20 per cent if the brewing industry moved to a more sustainable model of using reusable glass bottles, as it did in decades past.

And in the United States, according to a report published by the United States Chamber of Commerce Foundation in association with the Ellen MacArthur Foundation, *Achieving a Circular Economy: How the Private Sector Is Reimagining the Future of Business*, a move towards a circular economy could over 10 years generate an estimated US $1 trillion annually in economic value, create more than 100,000 new jobs and productively divert 100 million metric tons of waste – while also helping to negate the impact of rising commodity prices, which have wiped out long-term average manufacturing cost reduction (US Chamber of Commerce Foundation, 2015).

Circularity is arguably at its most efficient when parts are simply reused, without being melted down or otherwise reprocessed. In simple terms, this is the practice of **remanufacturing,** where end-of-life products are returned to the original manufacturer, stripped down or disassembled, with individual components being – where possible – refurbished and then once again reassembled into a complete product. At its starkest, remanufacturing is exactly what happens to the returnable milk bottle or beer bottle, but the concept is seen at its best in more sophisticated engineering and aerospace applications, often in the context of engine and gearbox manufacture.

Earth-moving equipment and heavy engineering manufacturer Caterpillar, for instance, maintains a circular economy sub-section of the sustainability section of its website, in which it describes how remanufacturing delivers lower prices for Caterpillar customers as well as delivering on their – and Caterpillar's – sustainability objectives. At the time of writing, the website points out that Caterpillar annually recycles over 150 million pounds of iron and steel, much of it in the form of end-of-life parts that have been refurbished and overhauled and returned to an as-new condition.

Writing in the United States Chamber of Commerce Foundation's 2015 publication *Achieving a Circular Economy: How the Private Sector Is Reimagining the Future of Business,* for instance, Bob Paternoga, general manager of Caterpillar Remanufacturing, explained how Caterpillar's remanufacturing business had become a global operation employing nearly 4,500 employees in 17 facilities, collectively remanufacturing engines, turbines, gearboxes, gas compressors, drivetrains, turbochargers and fuel systems, incentivizing customers with a pricing *system* that offers remanufactured parts at a significant discount to brand-new parts, and a pricing *structure* that includes a refundable deposit for the returnable 'core' item – the remanufactured engine, gearbox or whatever – that is to be remanufactured. Perhaps not surprisingly, reports Paternoga, core return rates were running at 94 per cent in 2014 (Paternoga, 2015).

Automotive manufacturer Renault is another company that has made a significant commitment to the circular economy and remanufacturing. Again making explicit reference to the circular economy on its website, the company points out that it is the only European car maker to have taken stakes in companies that recycle end-of-life vehicles.

Indra, Renault's joint venture with waste management firm Suez Environment, operates a network of 400 dismantlers that dismantled more than 95,000 vehicles in 2015. The resulting materials are centrally pooled and then – as at Caterpillar – sold at lower-than-new prices to end customers. Renault is also progressively setting up short-loop recycling systems to recycle raw materials such as steel, copper, textiles and polypropylene: over 30 per cent of the materials in its new vehicle output have been recycled, with a near-term target of increasing this to 40 per cent.

Moreover, the company's Choisy-le-Roi plant near Paris also builds major subassemblies from these recycled components, reckoning to produce 100 rebuilt engines for every 120 incoming unusable engines. In 2013, says Renault, the Choisy-le-Roi plant reconditioned and gave a second life to 25,370 engines, 15,930 gearboxes and 11,760 injection pumps – in the process, observes a McKinsey report, consuming 80 per cent less energy than normal manufacturing operations, 90 per cent less water and generating around 70 per cent less oil and detergent waste (Nguyen, Stuchtey and Zils, 2014). What's more, it adds, the Choisy-le-Roi plant delivers higher operating margins than Renault as a whole can boast.

Finally, away from the automotive industry, Dell Technologies (formerly Dell Computer Corporation) is another excellent example of remanufacturing in action, as the company's regular annual sustainability reports make clear. According to the 2016 report, for instance, plastics salvaged through Dell's 2,000 Dell Reconnect closed-loop recycling centres, located across the United States, found their way into 48 different Dell products, up from 19 the year before.

In all, 3.4 million pounds of plastics from old electronics products (of any brand) were combined with 10.7 million pounds of plastics from items such as CD cases and water bottles to create 14.1 million pounds of recycled plastic, up over 20 per cent from the previous year. In addition, by using this post-consumer recycled plastic instead of virgin plastic resin, Dell reckoned to have reduced greenhouse gas emissions by 9,790 metric tons of carbon dioxide equivalent (CO_2e) – the equivalent, said the company, of taking more than 2,000 passenger cars off the road.

Making a start

Even so, precisely what the individual business or individual product designer or design team can do about sustainability may not at first be obvious. What's more, from the standpoint of the individual product designer or the individual business for which he or she works, it may seem an overly grandiose aspiration – especially considering all the various ways in which humans interact with the environment and the various ways in which the planet's environment can be harmed.

Nor is this all. A decision to take environmental issues into consideration when coming up with a new or altered design is not one to be taken lightly, even assuming that a design can be made more environmentally friendly or sustainable. Quite rightly, product designers will have several concerns:

- Is it actually possible for a product's design to actually be made more environmentally friendly?
- Can an improved design really make a meaningful difference to the environment?
- Can a commercial case genuinely be made for a more sustainable design – in other words, whatever the design function wants to do, will the rest of the business, and senior management, support the move?

So let's consider the facts. Almost by definition, there is a strong linkage between human-made harm to the environment and economic activity. Yet at least 60 per cent of gross domestic product (GDP) is in the form of consumer expenditure – durable and non-durable products and services. In other words, there is an economic – and environmental – consequence to every visit that we make to the supermarket, and to practically every product.

As Brass, Bowden and Moseley observe (Brass *et al*, 2007), over 80 per cent of all product-related environmental impacts are determined by product design – and those impacts, especially when combined with the behaviour of consumers who may often be ignorant of them, are quickly amplified for example because:

- We don't use things efficiently.
  - Objects are often designed to become obsolete quickly. 98% of all products are thrown away within six months of purchase.
- We produce too much stuff, much of it toxic.
  - It is estimated that over 30 tonnes of waste are produced for every tonne of product that reaches the consumer in the UK.
- We don't dispose of things efficiently.
  - In the UK we produce about 100 million tonnes of mixed waste every year, most of which goes to landfill. This figure is growing by about 3% every year, one of the fastest European growth rates for waste.

Moreover, the vast majority of the products bought by consumers, businesses and governments are mass-produced, and sold by the million or billion. So in aggregate, at this scale, product design decisions in effect become amplified many times over, to the point where seemingly simple product choices at the design stage *do* make a discernible difference.

As an example, consider the progress made in reducing the wall thickness (and therefore weight) of the ubiquitous aluminium drinks can. When first in common use, in the 1950s, drinks cans weighed about 80 grams, a figure which had reduced to around 57 grams by 1970. By 1992 the weight was down to 16.5 grams, and 14.9 grams by 2001. Today a 330 millilitre aluminium can only weighs about 13 grams – less than a fifth of the weight in the 1950s. (The International Aluminium Institute, 2017; Australian Aluminium Council, 2018).

Each individual can, of course, saves just a few grams in weight. But, produced by the billion, these incremental reductions add up. And aluminium that *doesn't* go into cans is aluminium that doesn't have to be extracted, transported to the smelter (usually through a combination of rail and sea freight sometimes totalling thousands of miles), smelted (aluminium smelting is *extremely* energy-intensive), rolled and then transported through the downstream supply chain to the eventual end-customer.

Now throw in two more important components of GDP: the government's own expenditure on products, and the investments made by businesses and government in physical assets such as machinery and equipment. Added to the durable and non-durable products making up the consumer expenditure component of GDP, it's clear that a fairly high proportion of economic activity is in the form of buying products – products which of course have had to be designed. And so, for good or ill, those designs – in aggregate – will have a considerable bearing on the planet's environment, even though the impact of a single isolated product will be minuscule.

Small changes add up

Even so, is it possible for a single product's design – or its supply chain – to *really* make a meaningful impact on that product's environmental impact? Again, consider the facts. Broadly speaking, a product may be considered to have a beneficial impact on the environment if it:

- uses less of the Earth's resources to manufacture, pack, transport, or store;
- produces fewer or less harmful effluents, including carbon emissions, either in its manufacture, transport, or use;
- embodies innovations designed to make it easier to reuse, remanufacture, or recycle;

- actively incorporates recycled or remanufactured items in its design;

- substitutes renewable resources for non-renewable resources in its design – eg, paper instead of plastic;

- lasts longer and so postpones the need for replacement;

- contains components or raw materials designed to be sourced from local or 'near shore' sources of supply, rather than remotely located sources of supply – thereby reducing transport requirements and associated fuel consumption and pollutant emissions.

Put like this, the challenge seems more manageable. Not only are many of these things not inherently difficult – where technically possible – but the requirement is only to make improvements *at the margin*: that is, simply to design products that are more environmentally sustainable than their predecessors. Not *perfectly* environmentally sustainable – just *more* environmentally sustainable. Not at a punitive cost. Not in an impossible short timescale. And using today's technologies and today's production processes. In other words, the focus is not on making the product *perfect* from a sustainability point of view, but simply on delivering a product that is an improvement on what went before.

And, as ever, the sheer scale of mass production quickly ensures that small incremental improvements add up – especially when a number of those small incremental improvements are combined together in a single product. Consider computer and desktop printer manufacturer HP Inc (the computing hardware business created when IT giant Hewlett-Packard split in two in 2015), which has worked hard at sustainability, as its annual sustainability reports readily reveal. HP, these reports boldly declare, is 'reinventing how products are designed, manufactured, used, and revered as we shift our business model and operations toward a circular and low-carbon economy.'

According to the company's 2016 sustainability report, for instance (HP, Inc, 2016):

- 47% of new commercial HP desktop products contained over 10% of post-consumer recycled plastic content, up from 33% in 2014.

- 70% of HP commercial displays contained over 10% of post-consumer recycled plastics and 26% contained over 40% of post-consumer recycled plastics.

- The 3.4 billion HP ink and toner cartridges that the company had manufactured up to the end of 2016 contained over 88,900 tonnes of recycled content material – keeping 735 million cartridges, 70 million apparel hangers and 3.7 billion post-consumer plastic bottles out of landfills, instead recycling these materials for continued use.

- Over 80% of the company's ink cartridges contain 45–70% recycled content and 100% of its toner cartridges contain 10–33% recycled content. In 2016, HP used 9,000 tonnes of recycled plastics in HP toner and ink cartridges.

And what makes the company's commitment to sustainability all the more powerful is its scale, delivering 102 personal computers, 63 printers and 983 items of consumables to customers each minute of the day. Moreover, sustainability is embedded into the company's operations and practices. And suppliers, it declares, play an essential role in HP's circular economy strategy, leading the company to work closely with them to use materials, energy and water more efficiently, and to remove substances of concern from its products and manufacturing processes.

A journey of many steps

In short, take a company such as HP and study its sustainability reports over a period of time, and the picture that emerges is not of some grand sustainability vision, flawlessly executed in a single programme, but instead one of incremental improvement. The outcome: more sustainably designed – and more sustainably manufactured – products, generally achieved *one small step at a time*.

Is it possible to make a product lighter, or make it less wastefully? Then it is almost certainly more environmentally friendly. Does it produce fewer harmful effluents or pollutants – including carbon emission – either during manufacture, or transport, or in use? Then again, it is more environmentally friendly. Does it incorporate recycled or remanufactured products? Then again, it's a win for the environment. Does it use, say, paper or card in place of plastic? Ditto. And can it use straw-based packaging instead of paper or card? Even better. (Boxes and protective packaging made from wheat straw are similar in look and performance to corrugated paper-based equivalents, but the process to create them uses 40 per cent less energy and 90 per cent less water – and relies entirely on a waste product, straw, rather than an element of virgin-grown timber.)

Moreover, the focus is broader than simply the core product itself. These same considerations extend to the product's packaging, and the transport and distribution choices regarding how it reaches the end customer. A product that is lighter, for instance, is a product that is more fuel-efficient to transport – meaning less consumption of fuel and fewer emissions of pollutants such as carbon dioxide, sulphur and PM2.5 particulates. Likewise, a product that is packed more efficiently, so that more products can be fitted into a given cubic volume – a shipping container, perhaps, or a truck body – is again a product that delivers fuel savings and lower emissions levels.

Again – to stress – the focus is *not* on achieving perfection from a sustainability point of view at the first iteration. As with the practice of *kaizen* or continuous improvement in manufacturing operations, the trick is to view the journey as a series of incremental improvements. Not only does this make the process more manageable, but it is a way of viewing sustainability that is more likely to lead to a successful outcome; aiming for perfection at the first attempt is undeniably daunting.

CASE STUDY   Coca-Cola: 'rightsizing' containers saves plastic and reduces shipping costs

Sustainability and socially responsible practices have long been a focus of The Coca-Cola Company ('Coca-Cola'). In 1984, the company founded The Coca-Cola Foundation and pledged to use it to give back to the community 1 per cent of its previous year's operating income. While the Foundation's charitable objectives have evolved over time, women's empowerment, water sustainability and consumer well-being are recurring themes.

Water sustainability as a specific focus of Coca-Cola goes back to its partnership with the World Wide Fund for Nature, announced in 2007. This pursued five primary goals: conserving seven of the world's most important freshwater basins, improving water efficiency within Coca-Cola's own operations, reducing the company's carbon emissions, promoting sustainable agriculture and inspiring a global movement to conserve water.

In 2004, according to Coca-Cola's 2016 Sustainability Report (The Coca-Cola Company, 2017), the company was using 2.7 litres of water to make 1 litre of product – 'product' being the concentrate 'kit' of ingredients that Coca-Cola supplies to its primary customers, the network of global bottlers who actually manufacture and sell the cans and bottles of Coca-Cola drinks that consumers drink. Of this 2.7 litres, 1 litre of water was in the product itself and another 1.7 litres was used in the manufacturing process. As an indication of Coca-Cola's progress in water stewardship, by the end of 2016, that 2.7 litres had been reduced to 1.96 litres per litre of product, with a longstanding goal of reducing this further to 1.7 litres of water by 2020.

As well as reducing water consumption in its own direct operations, in 2008 Coca-Cola turned its attention to reducing water consumption within its supply chain and began to focus on the plastic containers that it purchased to ship ingredient kits to its bottler customers. These containers were usually purchased in sizes of 2.5, 5, 10, 20, 50 and 200 litres, and in some cases were despatched to the bottlers only 25 per cent full.

This excess 'head space' – or empty unused volume in each plastic container – then led to higher-than-necessary packaging costs, packaging waste and higher-than-necessary transport costs through poor vehicle and container utilization. In addition, plastic production is a highly water-intensive process, in which around 180 litres of water are consumed in order to produce 1 kg of plastic. With a typical 20-litre container weighing 1.2 kg, that meant that the transportation of 20 litres of concentrate product consumed 216 litres of water. Clearly, any reduction in plastic use would make an important contribution to Coca-Cola's overall water footprint.

A study into the feasibility of this took place within Coca-Cola's Americas operation, which embraced concentrate manufacturing operations in Puerto Rico and Atlanta (responsible for covering the United States and Canadian markets); Mexico (responsible for covering the Mexican market); and concentrate manufacturing operations in Costa Rica, Brazil, Argentina and Chile, which collectively focused on covering demand in Central and South America.

And in outline terms, it soon seemed possible to envisage that a reduction in plastic use of as much as 15 per cent could be achieved in the plastic packaging used in the downstream concentrate supply chain. This was made up of efficiencies in three particular areas:

- Using less plastic per container for the same container sizes, to be achieved through the use of thinner container walls or containers of different dimensions.
- Using different plastic container sizes, to decrease the amount of waste 'head space'.
- Redesigning the unit size of each concentrate 'kit', to make better use of container sizes.

Teams were set to work, exploring each area. One team would evaluate the feasibility of modifying existing kit configurations, in order to remove excessive head space. In parallel with this, a second team would work with suppliers to reduce the unit weight of existing containers. In the case of the third team, it was decided that in those cases where kit modification was not possible, additional container sizes were to be developed with suppliers, with the team balancing on a case-by-case basis the savings made versus the capital costs of new container moulds.

The first initiative – redesigning 'kits' to make better use of existing container sizes – was quickly successful. Some 60 per cent of existing kits were redesigned so as to ensure that head space would be at least 5 per cent, but less than 20 per cent.

The second initiative, to reduce the quantity of plastic per container, turned out to be more involved than had initially been envisaged. Because of the secret and proprietary nature of Coca-Cola's concentrate recipes, concentrate spills during transport are deemed to be one of the most serious operational issues that the business can face, as this could expose the concentrate composition to persons outside the Coca-Cola network. Instructions from headquarters were clear: spills do not happen in Coca-Cola's downstream concentrate supply chains.

Consequently, there were worries that thinner container walls might degrade the mechanical performance of a full plastic container, with internal calls for the development of lighter containers to be thoroughly documented, with proposed designs subjected to stringent testing. Eventually, agreements were reached with container suppliers whereby Coca-Cola shared the cost of the injection moulds used to create the containers, and suppliers in turn agreed to a lower price, subject to specified minimum order quantities being placed by Coca-Cola.

The third initiative was the one that took the longest to come to fruition. Several concentrate kits could not be redesigned around different standard container sizes, usually because the required production quantities at the bottlers were too small to justify larger kits. The search thus became one of identifying an optimum number of different packaging sizes in the range 2.5 to 20 litres, as well as a search for new container suppliers. Such suppliers not only had to possess – or develop – a controlled manufacturing environment approaching clean room conditions, but also be prepared to invest in a lengthy certification process for raw materials, in order to avoid any risk of flavour contamination. Finally, a new 7.5 litre container was approved.

Taken together, however, the three initiatives were an undoubted success. Coca-Cola's total shipments of containers within the Americas were reduced by 8 per cent, while container costs were reduced by 11 per cent over a four-year period – even though sales volumes had grown by an average of 3.5 per cent per year (The Coca-Cola Company, 2017).

The desiderata of sustainability

Put yet another way, from a more strategic perspective, a business with increased sustainability as an overarching goal will be trying to tick four sets of checklists both in terms of its own actions, and in terms of the actions it imposes upon customers and suppliers as a result of their mutual trading relationship:

1 From the perspective of the manufacture of individual products, and the materials from which they are manufactured, these objectives will be to:

 - use less;
 - throw away less;
 - recycle more;
 - reuse and refurbish more.

2 From the perspective of the performance of individual products, these objectives will be to design products that:

 - where applicable, last longer;
 - perform more efficiently, from an environmental perspective;
 - can be disposed of sustainably, or recycled sustainably.

3 From the perspective of the manufacturing operations in which these products are produced, these objectives will be to:

 - consume fewer resources (energy, water and consumables) during manufacture, shipping and disposal;
 - produce low levels of environmentally harming effluent and by-products;
 - achieve low levels of product and material scrap.

4 And finally, from the perspective of the downstream logistics and distribution processes that bring these individual products to their end customers, these objectives will be to:

 - minimize packaging, subject to product protection and safety requirements;
 - use packaging that has been sourced sustainably and can be disposed of (or recycled) sustainably;

  – maximize cubic utilization, thereby increasing transport efficiencies;
  – minimize weight, thereby also contributing to transport efficiencies.

In short, it's quite a list – and the good news is that product design, to a greater or lesser extent, either *can* – or *does* – have a bearing on all of these aspirations.

Will 'they' let us become greener?

Even so, many design functions will have a very prosaic concern: will their attempts to 'go green' be met with approval by senior management and others in the business? Put another way, is going green 'shareholder friendly'?

The answer is almost always an unqualified 'yes'. And for three main reasons – each relating, more or less directly, to improved competitiveness, or cost reductions, or both, as a result of developing more environmentally friendly sustainable products.

In fact, think for a moment about the objectives listed immediately above. Yes, they all have an impact on sustainability and environmental friendliness. But undeniably, many – if not most – of them also have a bearing on cost.

- Environmentally friendly products can be (and often are) *less expensive products* in terms of *direct material cost*: simply put, products that use fewer materials are products for which those materials don't need to be bought. Consider, once again, the reduction in the weight of a typical aluminium can – which as we have seen has come down significantly over the past 25 years or so. The impact on the environment is undoubted – but so too is the impact on the bottom line of the businesses buying those thinner and lighter cans.

- Environmentally friendly products are also products that have been manufactured more sustainably, consuming less energy, water and consumables, and emitting less effluent and unwanted by-products that businesses must pay to dispose of.

- Environmentally friendly products that are lighter, or packaged more efficiently in terms of cubic utilization or packaging material consumption, also help to deliver cost reductions downstream in terms of lower transportation costs, greater fuel efficiency, less packaging to throw away or recycle and fewer journeys.

- Finally, developing environmentally friendly products play to a business's corporate and social responsibility agendas. As such, it is viewed positively by customers, partners and employees. For marketing functions desperate for new and different promotional messaging, newly sustainable products are a gift from heaven.

In short, roll it all together and it's clear that businesses have an obvious incentive to pursue products that – either in manufacture, transport, or use – reflect a more sustainable agenda.

And, as we have seen, product design has a significant role to play in achieving that goal.

Checklist
Questions for design professionals

☐ How explicit are sustainability considerations in your product design processes and decision making? Is improved sustainability a formal goal? Does the broader business have sustainability targets or objectives?

☐ Does the product design function meet with the supply chain function to consider sustainability initiatives? If a supplier developed a more environmentally friendly offering in respect of existing products, how would the design function hear about it?

☐ Does your business maintain formal metrics on the sustainability of your products and production processes? (Think of HP's insights into the percentages of reclaimed plastic, for instance.)

☐ Have formal initiatives taken place to see if it is possible to improve the environmental performance of products in terms of cubic utilization, weight reduction, packaging reduction and the use of sustainably sourced packaging materials? If not, why not?

☐ What barriers would your product design processes encounter if attempts were made to improve product design sustainability? How might these be overcome?

Checklist
Questions for supply chain professionals

☐ How explicit are sustainability considerations in your supply chain sourcing and procurement decisions? Are suppliers measured or incentivized in relation to sustainability? And if not, why not?

☐ To what extent does sustainability feature in the operational reporting of your supply chain and logistics processes? Are formal metrics maintained covering such issues as recycling, fuel savings, carbon emissions and waste generation?

☐ Have formal initiatives taken place to see if it is possible to improve the cubic utilization and packaging performance of products?

☐ Would a relaxation of material specifications, tolerances and sourcing stipulations be advantageous from a sustainability point of view? If so, has this been raised with the product design function?

☐ Does the supply chain function meet with the product design function to consider sustainability initiatives? If a supplier developed a more environmentally friendly offering in respect of existing products, how would the design function hear about it?

References

2degrees (nd) ASDA Case Study, 2degrees Network [Online] available at: https://www.2degreesnetwork.com/2degrees/downloads/asda-case-study. pdf [accessed 29 January 2018]

Australian Aluminium Council (2018) *FAQs* [Online] available at: https://aluminium.org.au/faqs/ [accessed 1 February 2018]

Brass, C, Bowden, F and Moseley, J (2007) *Design for Sustainable Change: A positioning paper*, Design Council, London

Carson, R (1962) *Silent Spring*, Houghton Mifflin Company

Ellen MacArthur Foundation (2015) *Towards a Circular Economy: The business rationale for an accelerated transition* [Online] available at: https://www.ellenmacarthurfoundation.org/assets/downloads/ TCE_Ellen-MacArthur-Foundation_9-Dec-2015.pdf [accessed 30 January 2018]

GlaxoSmithKline (2015) Press Release, 2degreesnetwork [Online] available at: https://www.2degreesnetwork.com/2degrees/downloads/GSK_press_ release_-_14_Sept_2015.pdf [accessed 29 January 2018]

Hoskins, T (2017) H&M, Zara and Marks & Spencer Linked to Polluting Viscose Factories in Asia, *The Guardian*, 13 June [Online] available at: https://www.theguardian.com/sustainable-business/2017/jun/13/ hm-zara-marks-spencer-linked-polluting-viscose-factories-asia-fashion [accessed 30 January 2018]

HP, Inc (2016) *2016 Sustainability Report* [Online] available at: http://www8.hp.com/h20195/v2/GetPDF.aspx/c05507473.pdf [accessed 30 January 2018]

Kaye, L (2013) Clothing to Dye For: The textile sector must confront water risks, *The Guardian*, 12 August [Online] available at: https://www.theguardian.com/sustainable-business/dyeing-textile-sector-water-risks-adidas [accessed 30 January 2018]

Meadows, DH, Meadows, DL, Randers, J and Behrens III, WW (1972) *The Limits to Growth: A report for the Club of Rome's project on the predicament of mankind*, Universe Books, ISBN 0-87663-165-0

Nguyen, H, Stuchtey, M and Zils, M (2014) Remaking the industrial economy, *McKinsey Quarterly*, February [Online] available at: http://

thebusinessleadership.academy/wp-content/uploads/2016/03/Circular_economy.pdf [accessed 30 January 2018]

Paternoga, RK (2015) Caterpillar's remanufacturing business helps make sustainable progress possible, in *Achieving a Circular Economy: How the private sector is reimagining the future of business* [Online] available at: https://www.uschamberfoundation.org/sites/default/files/Circular%20Economy%20Best%20Practices.pdf [accessed 30 January 2018]

Pearce, DW and Turner, RK (1989) *Economics of Natural Resources and the Environment*, Johns Hopkins University Press, Baltimore, MD

The Coca-Cola Company (2017) *Water Stewardship: 2016 Sustainability Report* [Online] available at: http://www.coca-colacompany.com/stories/2016-water-stewardship [accessed 30 January 2018]

The Economist (2017) The Bad Earth: The most neglected threat to public health in China is toxic soil, 8 June, pp 24–6 [Online] available at: https://www.economist.com/news/briefing/21723128-and-fixing-it-will-be-hard-and-costly-most-neglected-threat-public-health-china [accessed 30 January 2018]

The International Aluminium Institute (2017) *Aluminium in Packaging – Lightweight* [Online] available at: http://packaging.world-aluminium.org/benefits/lightweight/ [accessed 30 January 2018]

US Chamber of Commerce Foundation (2015) *Achieving a Circular Economy: How the private sector is reimagining the future of business* [Online] available at: https://www.uschamberfoundation.org/sites/default/files/Circular%20Economy%20Best%20Practices.pdf [accessed 30 January 2018]

Further reading

Sigaria (2014) *Knowledge Share*, A Procurement Leaders 'Insight' paper

Building bridges 06
Making the change

Introduction

Fairly clearly, there are considerable business benefits to be derived through achieving greater cooperation and collaboration between the product design and supply chain functions – benefits as diverse as greater agility and responsiveness, greater sustainability, greater supply chain efficiency, lower supply chain risk and greater supply chain resilience.

But how exactly is this to be achieved? And how are businesses to know *when* it has been achieved, and *what* exactly has been delivered? This final chapter looks at how businesses can harness transformation methodologies such as **change management** and **business process re-engineering (BPR)** in order to achieve the sought-for cooperation and collaboration between the product design and supply chain functions. A final section of the chapter explores 'higher level' issues such as the role played by education, and the importance of good design being defined and celebrated in a manner that reflects the significance of supply chain considerations in design decisions.

The journey ahead

So here we are at the final chapter. Let's pause a moment to briefly review what we've learned and the journey that has taken us to this point.

- In Chapter 1, we looked at design 'classics' and design-led companies, at how product design can affect competitiveness and why the supply chain plays an important part in supporting the product design agenda.

- In Chapter 2, we looked in more detail at the interface between product design and the supply chain, and showed how building close links between the design function and the supply chain function can deliver significant benefits in terms of agility and responsiveness.

- In Chapter 3, we looked at the interaction between design decisions and both supply chain and design risk, and showed how making different choices can reduce a wide range of risks faced by businesses.

- In Chapter 4, we examined the role that the product design process plays in determining agility, and at how closer links with suppliers and customers, concurrent engineering and postponement strategies can help to deliver enhanced agility.

- In Chapter 5, we looked at the role that the product design process plays in improving businesses' sustainability performance, and at how designers can work with supply chain partners in order to improve sustainability.

In each chapter, we've looked at case study examples of real-world design and supply chain practices – both good and bad – and shown how different businesses have tackled the issues raised. We've also looked closely at leading design-led companies such as Zara owner Inditex, Apple, Dell and others, and highlighted the evident best practice to be seen there. Plus, of course, in each chapter we've posed 'checklist' questions for both design professionals and their supply chain colleagues to ask of themselves and their functions.

And finally, we've also described – in the '4C' model that we looked at in Chapter 2 and elsewhere – those characteristics of a desirable 'end state' – in other words, the key characteristics of an organization and its business processes where product design and the supply chain function are cooperating on a shared agenda aimed at such things as improving resilience and agility, reducing risk and improving sustainability performance.

Figure 6.1 The 4C model for the design-centric business

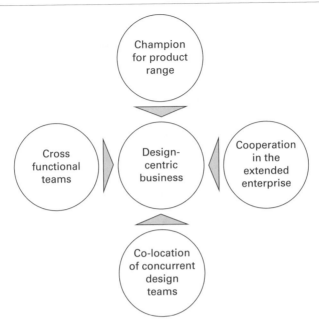

SOURCE Khan and Creazza (2009)

- **Cooperation in the extended enterprise,** ensuring that the impact of design decisions is understood by both the supply chain function and external suppliers. Through such cooperation, the business and its suppliers jointly mitigate design–supply chain related risks and ensure the smooth transition of products through the supply chain to the end customer through early supplier involvement.

- **Co-location of concurrent design teams,** with the design function and the relevant supply chain people being physically close together, in order to aid this close cooperation. All the functions that contribute to the design and development of a product are either physically co-located together – or if geographically dispersed, are *virtually* co-located with information transfer on as near real-time basis as possible – in order to ensure the smooth transition of products from drawing board to market.

- **Cross-functional multidisciplinary teams** of design professionals and supply chain professionals working concurrently, jointly contributing to the design process. These teams can also involve first- and second-tier suppliers, in order to benefit from early supplier involvement.

- **A product champion,** tasked with making sure that this dialogue takes place, and who is responsible for making the final call on its outcomes in terms of design decisions. The product champion manages the interface between design and supply chain functions, oversees the concurrent design process and ensures that there is a match between the product architecture and supply chain design.

A reasonable question, then, is: how to put all this into action? In other words, for a business that isn't *already* one where the product design and supply chain cooperate on a shared agenda of such things as improving resilience and agility, reducing risk and improving sustainability performance, then how can this be achieved?

Put another way, how can ordinary businesses – yours, perhaps – transform their product design and supply chain practices so as to put them in the same bracket as such luminaries as Apple, Dell, Inditex, Dyson and the rest?

That is what this final chapter, Chapter 6, attempts to outline. Be warned, though, that for those looking for a complete transformation – to become another Zara, for instance – there are few easy answers. To be sure, it's possible to make *improvements* quite straightforwardly: to many readers, for instance, the 'checklist' questions at the end of each chapter will have suggested issues to address, sometimes relatively easily. But a total transformation is another matter and requires commitment and effort right from the top of the organization.

Why? For one thing, as we saw in Chapter 2, product design and the supply chain function are two very different organizational silos. And in most businesses, organizational resistance can be expected – not least because separate silo agendas and reporting metrics will run counter to efforts to build cooperative bridges for the organizational good. For instance, if a consensus emerges to switch from sourcing from a low-cost overseas supplier to a more expensive domestic one, on the grounds of enhanced agility or improved supply chain resilience, then conventional reporting systems and accounting systems would regard this adversely, as a cost increase.

Costs can be measured and counted and can be built up to an overall product cost or variance, and ultimately to a bottom line. But agility or resilience are much more difficult to quantify. Agility at least can be captured in a positive sense: it is possible to measure a reduction in lead time, for instance, or possible to quantify the inventory impact of a postponement strategy. Resilience is a much, much trickier characteristic on which to get a handle: in the absence of something adverse occurring, resilience is simply a potential – an ability to recover more quickly than would otherwise be the case. Until actual supply chain disruption occurs, quantifying this recovery is next to impossible. And in the case of improved resilience in the form of making sure that something can't happen – taking sourcing away from a supplier located in an earthquake or tsunami zone, for instance – the problem is even greater. How do you measure the impact of an action to make sure that something can't happen?

Leadership from the very top will also often be required in order to get the commitment to undertake the necessary investment effort in reorganizing how things are done. Encouraging – or mandating – the product design and supply chain functions to work more closely together is no trivial matter: as we shall see, the phrase 'cultural revolution' may not be an inappropriate description. It will absorb management time, require an investment in reorganizing the design process, and may also require a physical relocation of people and functions where the design and supply chain functions are not presently co-located. In other words, there are costs to consider – even if only opportunity costs in terms of management time and attention – and most senior executives and boards of directors will rightly wonder as to the return on investment that is applicable.

Having wondered, though, they may be no wiser: while it might be possible to estimate *potential* outcomes – and here this book itself may be useful as a guide – it will be difficult to quantify these with any precision. At best, they may only be to indicate potential benefits: an aspiration, rather than a firm estimate grounded in actual expectation. Nor will the timings of those benefits necessarily be comfortable for organizations accustomed to thinking in terms of payback periods of two or three years. The benefits may flow, but they are likely to mount over time as experience builds, and greater interaction between the design and supply chain functions comes to feel more natural.

Moreover, there are risks to consider. Significant numbers of strategic corporate transformation projects fail – and if you're imagining that having the product design and supply chain functions share a common agenda isn't a strategic transformation project, think again. Put another way, if it were easy, more businesses would already be doing it, and this book very probably wouldn't have been written.

Why do strategic transformation projects fail? Experts in change management routinely identify any number of hurdles, ranging from botched execution to the absence of a 'burning platform' that focuses the entire organization on success. And again, while it's easily possible to characterize closer cooperation between the product design and supply chain functions as being highly desirable and even strategic, only in relatively rare cases will it be straightforward to see it as a survival-threatening 'burning platform'. Of the companies profiled in this book, for instance, only the challenges faced by Marks & Spencer in 1998 and 2003, and the production problems faced by Airbus with its Airbus 380 (both outlined in Chapter 2) probably come close to being a genuine burning platform.

In summary, then, however desirable it might be for businesses to achieve the product design and supply chain synergies so effortlessly demonstrated by some of the companies highlighted in this book, the truth is that there are significant challenges to overcome. Among others, organizational challenges, cultural challenges, justification challenges and risk challenges.

And yet, the prize remains worth striving for. If the lessons from companies such as Dell, Apple and Zara-owner Inditex tell us anything, it is that benefits of achieving greater product design and supply chain integration are significant, and strategic. Achieving greater product design and supply chain integration opens the door to delivering a transformation in performance of which most boards of directors can only dream. And with that transformation in performance achieved, brand dominance and value-enhancing returns will naturally tend to follow.

So how best, then, can this journey be undertaken? Let's take a look at a possible path to follow.

Defining the challenge

What do Apple, Dell and Inditex have in common? And what do ECCO, New Look, Dyson and Bodum all have in common with each other, as well as with Apple, Dell and Inditex?

The answer is straightforward: all of these design-centric businesses embraced the supply chain issues associated with their designs right from their earliest beginnings, when they were all start-ups. Put another way, it's not that they proceeded to operate conventionally and then changed: it's that they never operated conventionally in the first place and haven't changed subsequently, as they have grown.

And that provides a significant clue as to how other businesses – existing businesses that today are no longer start-ups – can emulate such success. In short, decades of organization-specific custom and practice must be thrown out, and new ways of working put in, in their place. In other words, it's not about *literally* being a start-up that matters, it's about throwing away the past, and starting again with a blank sheet of paper.

In particular:

- Alternative management processes must be defined, which codify and enshrine the closer cooperation between the product design and supply chain functions.

- Those alternative management processes must be successfully delivered and put in place, and previous methods of working discarded.

- Some kind of control framework and reporting mechanism must be established, so as to ensure that the closer cooperation between the product design and the supply chain functions is working as intended, and that design and supply chain decisions reflect the broader agenda that the business now wishes to pursue.

Put like that, the problem becomes more tractable. It becomes a process of business transformation – and as such, even if no magic wands or silver bullets are available, there is at least a significant body of management thinking and best practice to draw on. More than that, there is also a significant body of practical implementation expertise available, in business schools and management consultancies.

And so, recasting the challenge as one of business transformation – but business transformation firmly focused and targeted on greater product design and supply chain collaboration, rather than on, say, a leaner organization or a more customer-friendly order quotation process – it is possible to point to specific management approaches, which address the issues in question.

So let's quickly take a look at each of these in a little more detail, and then – in the next section of this chapter – flesh out how each might be harnessed to the challenge of bringing the product design and supply chain functions closer together.

Alternative management processes must be defined, which codify and enshrine the closer cooperation between the product design and supply chain functions.

This sounds very much like something that can be addressed by business process re-engineering, a business process definition methodology pioneered in the 1990s, and generally associated with Michael Hammer of the Massachusetts Institute of Technology. Popularly adopted by companies implementing ERP systems for the first time, the rationale was to use business process re-engineering to define and optimize business processes (and in the process strip out and eliminate unwanted or unnecessary activities), and then automate the defined process within an ERP system. More recently, the term 'business process management' has come into vogue, but for our purposes, it is precisely the re-engineering aspect of business process re-engineering which is attractive.

Those alternative management processes must be successfully delivered and put in place, and previous methods of working discarded.

Achieving this is the role played by change management within transformation projects. Change management recognizes that many large-scale changes within organizations actually fail (or at the very least, under-deliver) through either inertia, poor execution or active resistance, and aims to provide a methodology for successfully delivering such change.

Through actively engaging with the reasons for why change fails, and supporting the proposed change through programmes of education, cultural awareness, training, project management and effective leadership, the idea is not only to make the change process easier and more likely to be successful, but also 'sticky' – in other words, once change has been implemented, it remains.

Finally, it is worth noting the overlap with business process re-engineering and business process management, as these approaches often contain change management capabilities. That said, such capabilities are generally regarded as being less extensive than a full, formal change management programme.

Some kind of control framework and reporting mechanism must be established, so as to ensure that the closer cooperation between product design and the supply chain is working as intended, and that design and supply chain decisions reflect the broader agenda that the business now wishes to pursue.

There are various options here. Perhaps the simplest, though, is a combination of some basic metrics and the stage-gate review process that we looked at in Chapter 3.

The stage-gating would impose formal stages on the design process, at the end of which considering the various issues that we have been thinking about in this book must have been concluded.

Operationally, the process could be regarded as delivering similar results to the sort of thing seen in, say, Zara, by imposing a requirement for it to have taken place. And while not the same as building it into the culture of an organization, over time such reinforcing would tend to lead naturally to a cultural shift, in much the same way as often takes place when, say, organizations put in place initiatives to raise safety consciousness, or measures to promote equality.

'If you cannot measure it, you cannot improve it,' wrote Victorian physicist Lord Kelvin (Eveleigh, 2009). More recently, business managers rely on a pithier version of the same thing: *you get what you measure.*

So it seems sensible to incorporate into our approach some measurement methodology, so as to ensure that the organization is actually on track, and making progress towards its design and supply chain goals. These need not be complex and certainly don't need to be 'real time'. Management may care to display some of them on Business Intelligence-style digital dashboards, but this is probably overkill.

Making the change

So how, in detail, might each of these approaches be harnessed to the challenge of bringing the product design and supply chain functions closer together? Without being too prescriptive – much will vary according to the situation in which individual businesses find themselves – it is still possible to sketch out in some detail what businesses can do to effect the required transformation.

So let's take each of these three approaches in turn, and explore further how they can be used.

Re-engineering the design process

As we've seen, 80 per cent of a product's eventual supply chain costs are already present at the early stages of product design and development, meaning that design decisions can have significant through-life implications for risk, complexity and responsiveness.

That is why the focus needs to be on the design process, rather than the supply chain aspect of things. True, the supply chain function will have to effect changes as well, and do some things differently. But the main impetus for this will be the design process, as a redesigned process encourages – or rather, enforces – greater dialogue between the product design and supply chain functions, prompting greater reflection, greater understanding, and decisions regarding sourcing practice and policy.

As such, the idea is to codify the collaboration and cooperation that companies such as Zara have achieved through co-location of design and supply chain teams. Put another way, it is not the act of co-locating *itself* that delivers this collaboration and cooperation, but the fact that co-location makes it easier to conduct the necessary dialogue. But that dialogue has to take place in an environment where such interaction is encouraged: co-location does not of itself bring it about.

Business process re-engineering, as the name implies, is a management technique that aims to achieve this, which came to the fore in the early 1990s. And by leveraging business process re-engineering to re-engineer processes in which that interaction *is* encouraged, firms can achieve the benefits of greater interaction between the product design and supply chain functions not only in situations where the two functions *are* co-located, but also in situations where they are not.

In contrast to the way in which organizational processes are usually improved – by making changes to what is currently in place – business process re-engineering instead redesigns them 'from the ground up', ignoring whatever processes previously existed. The result is typically a greater level of improvement than can be obtained from incremental improvement, which can be regarded as a reflection of the extent to which many organizational processes contain once-necessary activities that are no longer required. It's also usually a quicker means of arriving at the desired end point.

The re-engineering process is deceptively simple. The starting point is to recognize that in any business, organizational functions are carrying out either processes or sub-processes within the business's overarching process – 'order to cash', for instance, in the case of a make-to-order manufacturer. But while these activities are central to the working of any business, they are not necessarily reflected in the way that a business or supply chain is actually structured.

Typically, adopting this process-centric view results in cutting across traditional organizational barriers: order acceptance, for example, may involve personnel from sales, finance, engineering and distribution. Another common observation is that within the 'traditional' organization, internal performance measures and objectives often prioritize narrow local concerns over those of the business as a whole. Again, a process-centric view eliminates this tendency.

Nor is there any need to develop this process view in isolation: models exist that can be used as building blocks. Many readers will be familiar with – or at least have heard of – the well-known Supply Chain Operations Reference (SCOR®) model, developed by the Supply Chain Council, an independent not-for-profit membership organization, in conjunction with management consulting firm PRTM, now part of consulting firm PricewaterhouseCoopers and AMR Research, now part of analyst firm Gartner.

Beginning with five so-called level one top-level processes – plan, source, make, deliver and return – each is then broken down into subsidiary processes, one of which is design. (A sixth level one top-level process, enable, was added in 2012; at the time of writing the most recent definition of the SCOR® model is SCOR®12.0, defined and published in 2017.) The Supply Chain Council is now a part of professional certification body APICS (formerly the American Production and Inventory Control Society), which provides training materials and guidance on using SCOR® (APICS, 2018).

The first stage of business process re-engineering is to consider the process's *mission*. Importantly, though, the idea is also to define it in terms of first principles: what is the process actually supposed to be delivering? Answering this question is not always straightforward, and can involve careful reflection. Clearly, for instance, the mission is to do more than just 'design a product'. Instead, businesses may want to phrase it in terms of 'designing products that delight customers', or, even better, 'rapidly designing products that delight customers'. It is worth stressing that careful thought is often required in order to define the present mission of the process, but that investment in time doing so usually pays dividends. In other words, the goal is to be able to answer the question: 'What are we actually trying to achieve at present?'

And that's because the next stage is to map – ideally visually – exactly how the present process works. More than just mapping the process, the idea is to match the mapped process to the mission, in order to assess its fitness for purpose. It could be, for instance, that the mapping reveals a process so convoluted and torturous that it is plainly apparent that in no sense can it deliver on a mission of 'rapidly designing products that delight customers'. Or, to hypothesize another mismatch, it could reveal a process that has no built-in review point at which designs are shown to customers or marketeers, making it difficult for designers to know with any certainty that their design will indeed delight customers.

Taken together, these first two stages are often called the 'As Is' phase, as the purpose is to pin down things as they presently are. This in itself can be extremely helpful: when Michael Hammer of the Massachusetts Institute of Technology – who is generally regarded as the originator of business process re-engineering – first described the approach in 1990, it was in a *Harvard Business Review* article entitled 'Reengineering Work: Don't Automate, Obliterate' (Hammer, 1990).

In other words, if the mapped process highlights activities taking place that don't add value, or that don't contribute to the mission, there is therefore a good argument for eliminating them. Equally, the 'As Is' phase also highlights, as we have just seen, sub-processes and activities that run counter to the mission, or which should exist, but don't. Again, the reader is referred to the SCOR® model: usefully, most treatments of SCOR® also include defined metrics for benchmarking purposes, as well as defined maturity levels, enabling businesses to calibrate themselves against best practice.

The third phase of business process re-engineering is to build a vision of the future – an envisioning exercise often called the 'To Be' phase. In other words, given a blank piece of paper, what could be achieved by redesigning the product design process differently?

Fairly clearly, with a question this broad, many things are possible. As we saw in Chapter 2, for instance, Marks & Spencer could have answered this question in terms of speed: compressing the time taken to bring a given product range to market, so as to more finely attune it to present fashions and market conditions. It almost goes without saying, too, that in attempting to answer a question like this, some sense of external perspective and challenge is almost always helpful.

The business may want to benchmark itself against the competition, for instance, in order to identify where its processes are market lagging, rather than market leading. Once again, using the SCOR® model for benchmarking or calibration purposes may reveal illuminating insights. The board or chief executive may want to set 'stretch' goals. Strategy consultants may be hired and customer focus groups conducted. In short, this is the time for bold, senior-level visions of what the business wants its new processes to look like, and how it wants them to operate.

In terms of the issues raised in this book, then, the mission would need rephrasing to reflect those product design-related goals that the business now wanted to reach. An existing mission of, say, 'rapidly designing products that delight customers' would need amending to reflect additional objectives in the areas of supply chain risk, sourcing, sustainability and responsiveness. As such, the revised mission might, for example, be something like 'agilely designing sustainable products that delight customers while being optimized for manufacturability, minimizing supply chain risk and meeting sourcing objectives in terms of cost, inventory holdings'.

The precise wording, and the precise set of objectives, will of course vary from business to business: as we saw in Chapter 2, there is a wide range of potential aspirations to choose from, including:

- design for rapid development;
- design for manufacture;
- design for product cost;
- design for supply;

- design for reduced supply chain risk;
- design for rapid time-to-market;
- design for sustainability;
- design for localization.

All of which, of course, will drive the interaction between the product design and supply chain functions in different directions – component reuse, postponement strategies, risk reduction, cost optimization, sustainability, or whatever else is appropriate. Nor will the focus necessarily be on one single objective in isolation: several goals may be followed at the same time, leading to the pursuit of several strategies simultaneously – component reuse, risk reduction and postponement strategies together, for instance. Each business is unique, and each business will need to make its own decisions about the nature of the interaction between the product design and supply chain functions that is most appropriate for its needs.

A final consideration in developing the 'To Be' phase of business process re-engineering is the need to be ambitious. Here, the emphasis is on the business's leadership, who must make sure that the re-engineered process is a sufficient break with the past so as to make a meaningful difference. Small, easy changes – while doubtless straightforward to effect, and ostensibly meeting the requirement of an apparently re-engineered process – will rarely deliver the required transformation. Instead, the business's leadership needs to demand changes significant enough so as to make a real difference.

Indeed, research by behavioural scientists Burke and Litwin has suggested that only an organization's leaders can create true 'transformational change'– change that is deep rooted, pervasive and long lasting (Burke and Litwin, 1992). Managers, in contrast, typically create 'transactional change': changes that are smaller and more short-term, and usually based on trade-offs within the organization, in which one group of people or managers agrees to do something in return for something else undertaken by another group of people or managers. Only a leader can genuinely impose change across organizational silos and across business functions – which, of course, is the very requirement when we are talking about building a re-engineered process that encourages greater interaction between the product design function and the supply chain function.

Reflecting on this point, here are two contrasting lists showing the roles of both the leader and the manager, adapted from Burke and Litwin's 'Leaders and managers transformational change versus transactional change' table.

Finally, with the mission defined and agreed, attention turns to the mapped process itself. The question to be resolved: with a new and different mission, what should the revised product design process look like? At what points in the design process are supply chain considerations to be taken into account? Where and how is this to happen? Ideally, co-location will make such interaction easier and more spontaneous. But the key point is that the re-engineered product design process calls for this interaction to happen – and happen in a manner consciously designed to be the most effective means of making that interaction as powerful as possible.

What form should this envisioned 'To Be' process take? Simplistically, the answer is some kind of flowchart, a diagrammatic representation of how designs should move through the newly envisaged process. But it is certainly possible to build upon this simplified representation, and businesses may well want to do that. Why?

Table 6.1 Leaders and managers transformational change versus transactional change

Leaders	Managers
Establish and communicate the vision	Carry out the vision
Motivate employees	Motivate, guide and direct employees
Establish and exhibit fundamental values	Translate fundamental values into business results
Focus on the future	Understand future direction but monitor the present

SOURCE Adapted from Burke and Litwin (1992)

Because when the co-location of design and supply chain people is a practical possibility, the actual mechanics of the interaction between them need not be tightly orchestrated. Insert an element of distance, though, and more consideration needs to be given to these mechanics.

One option is to use a *business process mapping language*, or – often in conjunction with a business process mapping language – a *business process mapping system*. Poirier and Walker, for instance, point out that while business processes have been around since the dawn of business itself, business process management languages and systems can be regarded as the next step in making them explicit, executable and adaptable. Such codification is also important if the interaction between the product design and supply chain functions is going to rely heavily on IT systems: the sort of modelling that a business process mapping system makes possible also serves to render the eventual transition to rules and stages within an IT system more straightforward than might otherwise be the case. Finally, a business process mapping system permits the use of simulations, so as to test and refine a business process (Poirier and Walker, 2005).

In summary, what the business ends up with is a redesigned set of business processes that formalize the touchpoints between the design function and the supply chain function. These should set out clear and visible business operating procedures for making sure that design considerations are taken into account by the supply chain function, and that supply chain considerations are taken into account by the product design function.

More than that, the procedures and processes should – if applicable – be hard-wired into the way that work is performed within the relevant enterprise systems running within the business. In a business with a product life cycle management (PLM) system, for instance, if the redesigned business process calls for the supply chain function to have visibility into each newly designed product as it reaches a certain point in its progress, then that visibility and sign-off should be hard-wired into the relevant workflow, so as to ensure that it happens.

Making the changes stick

Even so, many strategic change initiatives have failed. Many business process re-engineering exercises have failed. And common to many of these failures is an inadequate attention to **change management** – simply put, the art (or science) of making sure that proposed changes actually happen, and that once made, they remain in place. In other words, the previous methods of working must be discarded and those freshly designed alternative business processes must be successfully delivered and then sustained.

As we saw in Chapter 2, change management recognizes that many large-scale changes within organizations actually fail (or at the very least, under-deliver), through either inertia, poor execution or active resistance. To counter this tendency, change management aims to provide a methodology for successfully delivering such change.

And by first of all actively engaging with the reasons for why change fails, and then supporting the proposed change through programmes of education, cultural awareness, training, project management and effective leadership, change management strives not only to make the change process easier and more likely to be successful, but also 'sticky' – so that once change has been implemented, it remains.

That said, successful change management is difficult. Change management is not always successful. In the opinion of many, experience in successful change management makes an individual change manager more likely to succeed than any number of books, courses and theoretical articles. Nor is change management an exact science: most major consultancies have their own preferred 'flavour' of change management, and their own (often formalized) change management methodologies. Academics, too, debate and research which approaches to change management are more likely to be successful, and in which particular situations particular approaches to change management are most helpful.

Finally, as noted in the brief discourse on change management in Chapter 2, formal approaches to business process re-engineering and business process management often contain a change management element, not least because without those elements, early practitioners discovered that business process re-engineering's long-term success rate left something to be desired.

In short, businesses looking for guidance on change management do not have far to look. Compared to, say, the early 1990s, the subject is much better understood and has been much more extensively researched. Moreover, businesses themselves will have had much more exposure to change management principles, and be much more familiar with change management processes. Any business putting in place a radically different ERP or other enterprise IT system, for instance, is likely to have been exposed to change management processes.

That said, most approaches to change management exhibit a few common threads. They aim to *build a case for change*, for instance, with management jargon of 'burning platforms' and similar rhetorical devices. They aim to *ease the path to change* through training and communication exercises and describing a vision of how a different future could look. And (generally) they aim to *build on initial successes*, rather than rely on a 'big bang' approach to change that is effectively complete on day one.

The trick, if such there is, lies in the order in which these activities are carried out, and the effort invested in each one. For the individual reader in the context of his or her business, it seems reasonable to assume that truly effective change management probably involves some degree of customizing of published approaches to change management, tailoring them to the precise circumstances of the organization in question.

As we saw in Chapter 2, among those change management methodologies on offer, many observers point to the influential thinking of Harvard Business School's John Kotter, and the eight-step model outlined in his 1996 book *Leading Change*. Based on an earlier Harvard Business Review article entitled 'Leading Change: Why Transformation Efforts Fail' (Kotter, 1995), detailing the findings of a 10-year study of over 100 companies that attempted to effect significant changes, and which had quickly become the journal's best-selling reprint, *Leading Change* has undeniably stood the test of time, although the original eight-step model was revised slightly by

Kotter in 2014. A new edition of *Leading Change*, published in 2012, contains a new and retrospective preface from Kotter (Kotter, 2012).

Importantly, Kotter's model aids the change management process in two ways. First, it outlines a series of steps deemed necessary if major change is to take place – in other words, what has to happen. The value of this is that important steps don't get overlooked, or be regarded as peripheral: part of the formal process, management must undertake them, or run the risk of change failing. For management keen to simply roll up their sleeves and get stuck in, the need to first develop a vision may seem an encumbrance. And yet, suggests Kotter, having that vision – and communicating it – is important if change is to succeed. Second, Kotter's eight steps also place the required activities in an order of precedence deemed preferable if the chances of success are to be maximized. Put another way, the eight steps are, almost literally, a checklist. For management unused to accomplishing major changes successfully, such a checklist can prove vital.

Kotter's eight steps

1 Establish a sense of urgency.

2 Create a guiding coalition.

3 Develop a vision of a post-change state.

4 Communicate this change vision.

5 Empower others to act on the vision.

6 Generate short-term 'wins'.

7 Consolidate improvements, and produce still more change.

8 Institutionalize, or 'anchor', the new approaches.

A slightly different take on change management comes from another academic, Massachusetts Institute of Technology's Peter Senge, and was first outlined in a 1999 book, *The Dance of Change* (Senge *et al*, 1999). An advocate of systems thinking and organizational learning, change management to Senge is less about developing an initial vision, and more about leveraging small-scale changes and pilot projects so as to build the case for change.

Put another way, while Kotter pre-supposes a situation in which change is required or even essential – the 'burning platform' – and then suggests how that change can be sold and implemented, the Senge view of change is more gradual, with an emphasis on establishing what works, and trying to do more of it.

The Senge approach can be summarized as (Henley Forum, nd):

- Start small.
- Keep goals realistic.
- Grow steadily.
- Don't attempt to plan the whole thing.
- Stay close to ongoing change efforts.
- Expect challenges and mis-steps.
- Reinforce activities that are going in the right direction.
- Be open to feedback.

Which particular change management methodology or approach to adopt? The reader is hardly short of choices, and much will depend on the circumstances of the individual organization. In terms of the choice between the Kotter and Senge approaches, for instance, the Kotter approach might be considered a way forward for achieving *significant* change – an organization-wide strategic change initiative, perhaps – while Senge's less prescriptive approach might be considered more suitable for smaller-scale change.

But – to stress the point again – while these are two well-known approaches, they are two of many. For the reader wanting to learn more about change management before making a commitment to a particular methodology, one particularly recommended book is Esther Cameron and Mike Green's excellent *Making Sense of Change Management: A Complete Guide to the Models, Tools and Techniques of Organisational Change* (Cameron and Green, 2015).

The important thing, as management attempts to bring together two functions previously regarded as disparate, and pursuing two very different agendas, is that there *is* a change management methodology in place, so as to provide the newly defined business processes with the best possible chance of being genuinely adopted and delivering the sought-for changes in behaviour.

For if there is one overarching message from the present book it is that there are significant gains to be had from the supply chain and product design functions working more closely together, and that an investment in helping this to happen is almost always going to be worthwhile. Don't make the mistake of thinking that the required level of cooperation and shared thinking can be achieved without an investment in appropriate change management: if achieving it was easy, then the required level of cooperation and shared thinking would be more readily observable. And, as we have amply seen, it isn't.

Staying the course

Finally, with the design and supply chain functions now working more closely together, it seems sensible to put in place some kind of *control framework* and *reporting mechanism*.

The purpose of the control framework is simple: to provide a form of management governance, in order to make sure that the new business processes are being complied with, and that both product design decisions and product design-related supply chain decisions formally acknowledge the need for the requisite cooperation and discussion to have taken place.

Likewise, the purpose of the reporting mechanism is equally straightforward. What gains are being achieved by formally requiring the product design and supply chain functions to work together? To what extent *are* they working together? What progress is being made towards such objectives as, say, improved sustainability? How much more resilient are the business's supply chains?

In both cases, there are various options available. And inevitably, given the organization-specific nature of both control frameworks and reporting metrics, by necessity this portion of this chapter is less prescriptive. A control framework that works well for one organization might not work for another, and a reporting metric that provided insight to one organization might not be relevant to another. Inevitably, businesses will have to choose approaches that work for them, and which are consistent with their own organization structure, management style, culture and values. That said, in terms of the control framework, the observation has already been made that the stage-gate review process that we looked at in Chapter 3 could well deliver value.

The stage-gating would impose formal stages on the design process, at the end of which any consideration of the various issues that we have been thinking about in this book must have been concluded. Stage-gating is conventionally used in the context of product development decisions for 'go/no-go' decisions: not letting the development process proceed any further before a number of fundamental questions have been satisfactorily answered. Is there a market for the product? Is the market large enough for a developed product to be sold profitably? Does the product actually work? Does it meet cost targets? Can it be developed in an acceptable time? And so on, and so on.

It is not too big a stretch, surely, for this process also to be extended to a consideration of the joint product development and supply chain issues that we have seen in this book. A number of the checklist questions at the end of each chapter could easily be adapted for this purpose. So too could some of the bulleted discussion points that we have covered in a number of chapters – Chapter 5, on sustainability, for instance, is rich in such discussions.

And irrespective of the underlying detail of the stage-gate considerations in question, the key point is that the stage-gating process enforces a pause for review, and a formal means for the organization to ask itself whether it has met its own aspirations in terms of the relationship that it envisages between the supply chain and product design functions. Quite apart from the precise wording of the stage-gate tests to be applied, the very process of pausing to consider if the benefits of that closer relationship have been fully realized is a powerful and formal reminder that those benefits exist, and are there to be had.

Finally, businesses will want to supplement this stage-gating process with a set of metrics to illuminate the enhanced cooperation between their product design and supply chain functions. Once again, individual businesses will have different requirements in this respect, but conceptually, such metrics will serve two important functions:

- First, businesses will want to highlight the progress that they are making in developing the requisite enhanced cooperation between their product design and supply chain functions.

- Second, businesses will want to see metrics that highlight the benefits that such cooperation has delivered, in the form of more resilient supply chains, more sustainable products, more agilely-developed products and so on.

Conceptually, too, the nature of these metrics – and their mix – will also change over time. To begin with, businesses will probably want to see more metrics focusing on the progress that they are making in having their product design and supply chain functions working more closely together. Their cooperation will be relatively recent, new business processes may need to 'bed in' and there will in any case be relatively few benefits to report on. Over time, though, when cooperation between the two functions becomes more in the nature of 'business as usual', it will be natural for businesses to need less reassurance on the extent and nature of the cooperation between the two functions, and instead want to see a greater focus on what exactly that enhanced cooperation is delivering.

Whatever the precise mix of metrics, they need not be especially complex, and certainly don't need to be 'real time'. How they are presented will to some extent depend on the management and reporting culture of the individual organization, but it seems reasonable to assume that inclusion within regular board-level reporting packs or regular executive management committee reporting packs will suffice. As noted earlier, Business Intelligence-style digital dashboards are probably overkill.

Which specific metrics or reporting frameworks should businesses use? Again, no two businesses are the same and it is difficult to be prescriptive. But in outline, and at a high level, the following measures are probably a good starting point for thinking about the issue.

Measures to highlight the growing cooperation between the product design and supply chain functions:

- What percentage of newly launched products have been the subject of cooperation and collaboration between the product design and supply chain functions?
- What percentage of the overall product range have been the subject of such cooperation and collaboration?
- Are the two functions co-located? If not, how are plans progressing to achieve this?
- Are they electronically linked through workflow? If not, how are plans progressing to achieve this?
- How many joint meetings have taken place? Monthly? Quarterly? Annually?

Measures to highlight the benefits of that cooperation:

- How agile are product design processes? How is that agility improving?
- How many individual design changes have been made as a result of cooperation and collaboration between the product design and supply chain functions? Monthly? Quarterly? Annually?
- How are improvements being made to products' sustainability as a result of that cooperation and collaboration? What percentage of products incorporate recycled or recyclable materials? How quickly are these measures improving?

- How many individual sourcing changes have been made as a result of cooperation and collaboration between the product design and supply chain functions? Monthly? Quarterly? Annually?
- What other supply chain risks have been identified and eliminated – or reduced – as a result of that cooperation and collaboration?

The bigger picture

Finally, let's briefly turn to consider the broader role of design, design education and the business environment in which cooperation and collaboration between the design function and the supply chain function takes place.

To begin with, it will be clear to the reader who has read the preceding chapters in this book – and who has mulled over the various challenges raised in the present chapter – that an obvious question is why this cooperation and collaboration does not take place already. This, it turns out, is a question with several answers.

We need industrial designers, not just designers

First, as we saw in Chapter 1, the product design process, and product design education, often tends to point designers towards aesthetic considerations and design proliferation rather than a sober consideration of what their design choices mean in supply chain terms. To this end, as we again remarked in Chapter 1, there is a strong case for preferring the words 'industrial design' and 'industrial designer' to 'design' and 'designer'. While not unknown in the UK and Europe, these terms are far more commonly used in the United States, where industrial design is promoted and promulgated through the Industrial Designers Society of America.

At the very least, a practice of recruiting designers as industrial designers, and appointing people as industrial designers with the words 'industrial designer' in their job titles, would be a sensible first step to seeing designers more open to considering the link between the design decisions that they take and the supply chain implications of those decisions.

Put another way, while good design is rightly celebrated, it is not clear – even after pivotal publications such as the 2005 *Cox Review* – that the concept of good design adequately embraces industrial efficiency in the form of supply chains that are efficient and resilient. To this end, it would be good to see bodies such as the UK's Design Council, and its equivalents around the world, do more to promote this notion (Cox, 2005).

Skills gap

Second, there seems little doubt that the education system is contributing to this observed lack of cooperation and collaboration between the design and supply chain functions. That said, educators will no doubt retort that their curriculums, courses and qualifications are predicated upon what their end students are deemed to be demanding, with those demands in turn being predicated upon what employers are supposed to be looking for in new recruits.

Consider the facts. In many industries – and certainly industries such as footwear, furniture, fashion and general apparel – designers are in effect creative artists, working with materials such as fabric, wood or leather to create stylish objects that will visually impress. In still other industries, creatively inspired stylish designs are wrought from aluminium, stainless steel or glass, to again deliver commercial success in a tough competitive marketplace. As examples, think of Sir James Dyson's eponymous products, or Steve Jobs' and Sir Jonathan Ive's extensive output at Apple, or Peter Bodum's stylish tableware.

Simply put, in this view of product design, the designer is more than likely to have gone to art school or fashion design school, and to very likely consider themselves to be a creative artist. And not, to put it bluntly, as someone with either an interest or a working knowledge of the intricacies or principles of supply chain management.

There's another view of product design, of course, where product designers will have trodden a different educational path. In many industries, design is shared between creative stylists, who produce or influence a product's exterior appearance, and technical specialists – engineers of some sort, generally – who are responsible for the design of what lies under or within that exterior appearance. Electronic and electrical products, for instance, or motor vehicles, are examples of such products.

And here, the bulk of the product design workload, and the design and selection of the majority of the product's components, will be in the hands of electronic engineers, materials engineers, electrical engineers, aerospace engineers or mechanical engineers. They too – quite rightly – consider these skill sets to be their primary specialism and will rarely, if ever, lay claim to any deep insights into the principles of supply chain management.

Finally, too, supply chain professionals have what might be considered as the opposite problem: they know about supply chain management, but might have a limited grasp of aesthetic principles, or understand why a designer wants a particular material, fabric or colour. Which is why cooperation and collaboration between the design and supply chain functions is so important: in essence, designers go to art school or engineering college and supply chain professionals go to business school. They lack each other's professional insights and language and are too centred on the demands of their profession and their internal function within the organization.

What to do about this? In part, it is the responsibility of educators to broaden their curriculums, or offer niche specialisms within those curriculums that include the required content. A master's degree in supply chain management with special reference to the textile industry, for instance, or the motor industry or the aerospace industry. Slowly, such courses are beginning to emerge, but the pace could be significantly faster. Design courses, too, could include an element of supply chain management – something that again would be accelerated if 'design' became '*industrial* design'. Likewise with engineering courses. There is little point in continuing to bring on generations of engineers who will go on to design products that are substandard from a supply chain point of view. And yet, if nothing changes, that will be very largely what happens.

What might change this gloomy prognosis, if educators don't move quickly enough or effectively enough? Well, employers could take a stand and begin to explicitly require such a fusion of skills, by using skills specifications in job advertisements to send a signal to the marketplace about the skills and the education that they value. If the experience of other emerging skill sets is anything to go by, such messages quickly get heard.

Celebrate the fusion

Finally, industry, government and the relevant professional bodies can add their voices, by placing a higher degree of emphasis on the interaction between product design and the supply chain in how they shape, influence and celebrate good design.

Few of the studies and reviews referred to in Chapter 1, for instance, make much reference to the supply chain aspect of design. Generally, the discussion stops at design for manufacturability. Design for agility? Design for supply chain resilience? Design for inventory efficiency? Their voices are largely silent.

And yet, it can be done – as we see with design for sustainability, where plaudits abound. The moral is clear: the headlines and glittering prizes can be there, if the will to celebrate supply chain success is there. And in that, every reader of this book can play a part.

Checklist

Questions for design professionals

☐ What skill sets does the product design function look for when recruiting? Should these be changed, or added to?

☐ Should design function professionals receive any training or education on supply chain issues?

☐ What barriers exist to the change management and business process re-engineering processes described in this chapter? How difficult might they be to overcome?

☐ Do product designers' job titles and job descriptions hinder or reinforce their willingness and ability to cooperate and collaborate with the supply chain function?

☐ As a result of reading this book, have your perceptions of the benefits of greater cooperation and collaboration between the product design and supply chain functions changed? How?

Checklist
Questions for supply chain professionals

☐ To what extent do people working in supply chain functions such as procurement and sourcing understand the agendas of their product design colleagues? How might this be improved further?

☐ What barriers exist to the change management and business process re-engineering processes described in this chapter? How difficult might they be to overcome?

☐ If the product design and supply chain functions are not presently co-located, how could this be achieved? How realistic is electronic workflow as an alternative? What would the business case to achieve this look like?

☐ Do supply chain professionals' job titles and job descriptions hinder or reinforce their willingness and ability to cooperate and collaborate with the product design function?

☐ As a result of reading this book, have your perceptions of the benefits of greater cooperation and collaboration between the product design and supply chain functions changed? How?

References

APICS (2018) SCOR – The Most Recognized Supply Chain Framework [Online] available at: http://www.apics.org/apics-for-business/products-and-services/apics-scc-frameworks/scor [accessed 31 January 2018]

Burke, WW and Litwin, GH (1992) A Causal Model of Organizational Performance and Change, *Journal of Management*, **18** (3), 1 September, pp 523–45

Cameron, E and Green, M (2015) *Making Sense of Change Management: A complete guide to the models, tools and techniques of organisational change*, 4th edn, Kogan Page, London

Cox, G (2005) *Cox Review of Creativity in Business: Building on the UK's strengths* [pdf], HMSO [Online] available at: http://webarchive.nationalarchives.gov.uk/20120704143146/http://www.hm-treasury.gov.uk/d/Cox_review-foreword-definition-terms-exec-summary.pdf [accessed 31 January 2018]

Eveleigh, C (2009) If you cannot measure it, you cannot improve it, *Getting to Excellent* [Online] available at: http://www.gettingtoexcellent.com/2009/05/if-you-cannot-measure-it-you-cannot.html [accessed 31 January 2018]

Hammer, M (1990) Reengineering Work: Don't automate, obliterate, *Harvard Business Review*, July

Henley Forum (nd) *Henley Forum - Knowledge in action leaflets* [Online] available at: https://www.henley.ac.uk/research/research-centres/the-henley-forum-for-organisational-learning-and-knowledge-strategies/folks-knowledge-in-action [accessed 31 January 2018]

Khan, O and Creazza, A (2009) Managing the Product Design–Supply Chain Interface: Towards a roadmap to the design centric business, *International Journal of Physical Distribution and Logistics Management*, **39** (4), pp 301–19

Kotter, J (1995) Leading Change: Why transformation efforts fail, *Harvard Business Review*, March–April, **73** (2) pp 59–67, (republished January 2007)

Kotter, J (2012) *Leading Change*, Harvard Business Review Press, Brighton, MA

Poirier, C and Walker, I (2005) *Business Process Management Applied: Creating the value-managed enterprise*, J Ross Publishing, Boca Raton, FL

Senge, P, Kleiner, A, Roberts, C, Ross, R, Roth, G and Smith, B (1999) *The Dance of Change: The challenges to sustaining momentum in a learning organization*, Nicholas Brealey Publishing, London

Further reading

Johansson, H and Carr, D (1995) *Best Practices in Reengineering: What works and what doesn't in the reengineering process*, McGraw-Hill, London

APPENDIX

New product design and development risks: the Boeing 787 case

CHRISTOPHER S TANG[1]

August 15, 2016

Abstract: To compete for market share, Boeing decided to develop the 787 Dreamliner by creating new value for its customers. The 787 Dreamliner is not only a revolutionary aircraft using advanced technology, the design of its supply chain is intended to drastically reduce development cost and time. However, despite significant management effort and capital investment, Boeing faced a series of delays in its schedule for the maiden flight and plane delivery to customers. This paper analyses Boeing's rationale for the 787's unconventional supply chain, describes Boeing's challenges for managing this supply chain and highlights some key lessons for other manufacturers to consider when designing their supply chains for new product development.

Key words: New product development, product design, supply chain risks

Introduction

Boeing has been the dominant manufacturer of commercial aircraft and it was a shock when Boeing's market share dropped below that of Airbus (owned by EADS) for the first time in the late 1990s. After wrestling between reducing the costs of existing types of aircraft and developing a new aircraft to raise revenues through value creation, Boeing convinced the board to develop an innovative aircraft, the 787 Dreamliner, quickly (in four years instead of six) and cheaply (US $6 billion instead of US $10 billion) in 2003. (Throughout this chapter, we shall use the terms '787 Dreamliner', '787' and 'Dreamliner' interchangeably.)

To compete for market share, Boeing's value creation strategy for passengers was to improve their travel experience through redesign of the aircraft and offer significant improvements in comfort. For instance, relative to other aircraft, over 50 per cent of the primary structure of the 787 aircraft (including the fuselage and wing) is made of composite materials (Hawk, 2005). As compared to the traditional material (aluminium) used in aircraft manufacturing, the composite material allows for increased humidity and pressure to be maintained in the passenger cabin, offering substantial improvement to the flying experience. Also, the lightweight composite materials enable the Dreamliner to take long-haul flights. Consequently, the Dreamliner allows airlines to offer direct/non-stop flights between any pair of cities without layovers, which is preferred by most international travellers (Hucko, 2007). Table A.1 and Figure A.1 compare the 787 aircraft with other popular aircraft.

Boeing's value creation strategy for its key immediate customers (the airlines) and its end customers (the passengers) was to improve flight operational efficiency by providing big-jet ranges to midsize aircraft while flying at approximately the same speed (Mach 0.85).[2] This efficiency would allow airlines to offer economical non-stop flights to and from more and smaller cities. In addition, with a capacity between 210 and 330 passengers and a range of up to 8,500

Table A.1   Comparison of selected Boeing and Airbus aircraft

Airline family	Max range (nautical miles)	Max capacity (passengers)[1]	Empty weight (lbs)	Cruising speed (mph)	Operations strategy
737–800	3,000	189	91,000	514	Direct flights to multiple cities
747–8	8,000	467	410,000	570	Hub to hub
787–9	8,500	330	254,000	561	Direct flights to multiple cities
A380–800	8,200	555	610,000	561	Hub to hub

1 Measured in terms of typical seat configuration. For example, the total number of seats can be higher if more space is allocated for the economy-class cabin and less space for the first- and business-class cabins.

Figure A.1   Dreamliner and A380 size comparison

Up, Up and Away

Weighing in at 280 metric tons and with a wingspan as wide as a football field is long, the Airbus A380 is the world's largest passenger jet, designed to carry about 850 passengers between hub airports. In contrast, Boeing says its smaller, fuel-efficient 787 Dreamliner will allow for direct flights between more cities even at great distances.

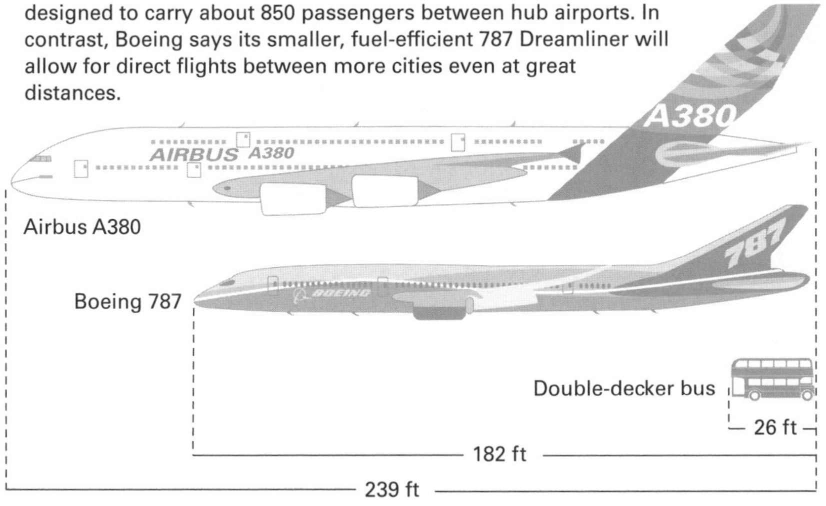

AIRBUS A380

Airbus A380

Boeing 787

Double-decker bus

26 ft

182 ft

239 ft

SOURCE Airbus, Boeing

Table A.2 Dreamliner features with benefits for airlines and passengers

Feature	Values to airlines (immediate customers)	Value to passengers (end customers)
Composite material	– Faster cruising speed, which enables city-pair non-stop flights – Fuel efficiency (lighter material lowers operating cost) – Corrosion resistance (lower maintenance cost) – Stronger components that require fewer fasteners (lower manufacturing cost)	– Faster cruising speed, which enables city-pair non-stop flights – Higher humidity in the air is allowed, which increases comfort level
Modular design that allows for two types of engines (General Electric GEnx and Rolls-Royce Trent 1000)	– Flexibility to respond to future circumstances (market demand) at a reduced cost – Simplicity in design allows for rapid engine changeover	– Cost savings with cheaper and faster engine changeover may be passed on to passengers
Large and light sensitive windows	– Lower operating costs due to less need for interior lighting	– 'Smart glass' window panels work like transition lenses, controlling the amount of light automatically, decreasing glare and increasing comfort and convenience
Redesigned chevron engine nozzle (serrated edges)	– Reduction in community noise levels	– Reduction in interior cabin decibel level
Easy preventive maintenance	– Boeing provides service so planes are in operation for longer periods of time	– Fewer delays due to mechanical problems

nautical miles, the 787 Dreamliner is designed to use 20 per cent less fuel for comparable missions than today's similarly sized aircraft. The cost-per-seat mile is expected to be 10 per cent lower than for any other aircraft. Also, unlike the traditional aluminium fuselages that tend to corrode and fatigue, 787's fuselages are based on the composite materials, which reduce airlines' maintenance and replacement costs (Murray, 2007). Table A.2 provides a summary of the Dreamliner's benefits for both the airlines and their passengers.

Due to the unique value that the 787 provides to the airlines and their passengers, Dreamliner was the fastest-selling plane in aviation history. The Dreamliner programme has been considered a model endeavour, combining novel technology and production strategies. As of 16 November 2008, Boeing (www.boeing.com) had received orders from more than 50 airlines that accounted for 895 Dreamliners. The overwhelming response from the airline industry about Boeing's 787 forced Airbus to quickly redesign its competitive wide-bodied jet, the A350, to make it even wider, which was later re-released as the A350XWB for 'extra wide body' (Wallace, 2006a).

Besides sales, the stock market responded favourably when Boeing launched its 'game-changing' 787 Dreamliner programme in 2003. As shown in Figure A.2, between 2003 and 2007, Boeing's stock price increased from around US $30 to slightly over US $100. Boeing announced a series of delays beginning in late 2007 and the market has reacted negatively (Figure A.2). The negative market response is somewhat expected as publicity of Boeing's supply chain issues become increasingly evident. As shown in Figure A.2, Airbus shared a similar fate after announcing a series of delays for the delivery of its A380 in early 2006 (Raman *et al*, 2008). Despite significant capital investment and management effort, Boeing is currently facing continual delays (for more than two years) in its schedule for the maiden flight and plane delivery to customers as of this writing (Sanders, 2009c). After numerous failed attempts to get its 787's composite rear fuselage supplier back on track, Boeing finally decided to acquire Vought's South Carolina facility at a cost of US $1 billion on 8 July 2009 (Sanders, 2009b). This observation motivates us to examine the underlying causes of Boeing's challenges in managing 787's delivery schedule.

Figure A.2   Historical stock prices of Boeing and Airbus compared to the S&P500

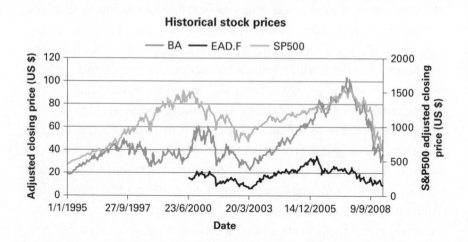

In this chapter, we shall examine Boeing's rationale for the 787's unconventional supply chain in the next section. We will then present our analysis of the underlying risks associated with its supply chain. In later sections, we describe Boeing's risk mitigation strategies to expedite its development and production process. The penultimate section highlights some key lessons for other manufacturers to consider when designing their supply chains for new product development. We conclude this chapter with an epilogue.

The design of 787 Dreamliner's supply chain

To reduce the 787's development time from six to four years and development cost from US $10 billion to US $6 billion, Boeing decided to develop and produce the Dreamliner by using a new and unconventional supply chain in the aircraft manufacturing industry. To reduce development time and cost, the 787's supply chain was envisioned to keep manufacturing and assembly costs low, while spreading the financial risks of development to Boeing's suppliers. Unlike the 737's supply chain that requires Boeing to play the traditional role of a key manufacturer who assembles different parts and subsystems produced by thousands of suppliers (Figure A.3), the 787's supply chain is based on a tiered structure that would allow Boeing to foster strategic partnerships with approximately 50 tier-1 strategic partners. These strategic partners serve as 'integrators' who assemble different parts and subsystems produced by tier-2 suppliers (Figure A.4). The 787 supply chain depicted in Figure A.4 resembles

Figure A.3   A traditional supply chain for aircraft manufacturing

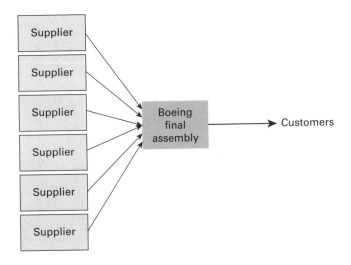

Figure A.4   Redesigned supply chain for the Dreamliner programme

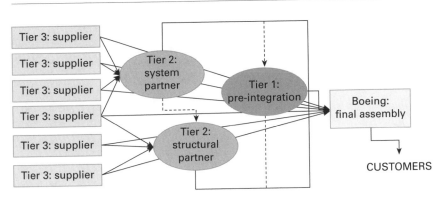

SOURCE adapted from Steve Georgevitch, Supply Chain Management, Boeing Integrated Defense Systems

Figure A.5 Dreamliner subassembly plan

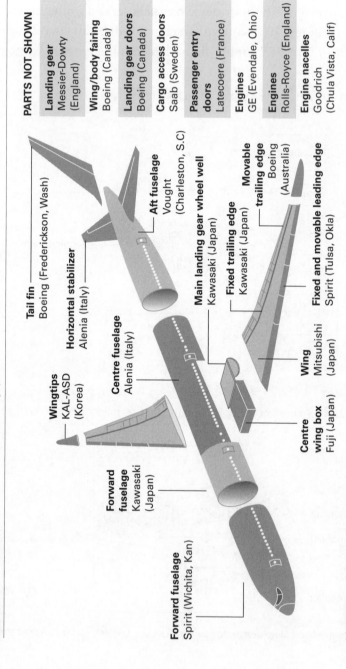

PARTS NOT SHOWN

Landing gear
Messier-Dowty
(England)

Wing/body fairing
Boeing (Canada)

Landing gear doors
Boeing (Canada)

Cargo access doors
Saab (Sweden)

Passenger entry doors
Latecoere (France)

Engines
GE (Evendale, Ohio)

Engines
Rolls-Royce (England)

Engine nacelles
Goodrich
(Chula Vista, Calif)

Tail fin
Boeing (Frederickson, Wash)

Horizontal stabilizer
Alenia (Italy)

Aft fuselage
Vought
(Charleston, S.C)

Wingtips
KAL-ASD
(Korea)

Centre fuselage
Alenia (Italy)

Main landing gear wheel well
Kawasaki (Japan)

Movable trailing edge
Boeing
(Australia)

Fixed trailing edge
Kawasaki (Japan)

Forward fuselage
Kawasaki
(Japan)

Fixed and movable leading edge
Spirit (Tulsa, Okla)

Wing
Mitsubishi
(Japan)

Centre wing box
Fuji (Japan)

Forward fuselage
Spirit (Wichita, Kan)

SOURCE *Seattle Post-Intelligencer,* Seattle

Toyota's supply chain, which has enabled Toyota to develop new cars with shorter development cycle time and lower development cost (Tang, 1999). Table A.3 highlights the key differences between the 737's supply chain and the unconventional 787 supply chain. For instance, under the 787's supply chain structure, these tier-1 strategic partners are responsible for delivering complete sections of the aircraft to Boeing, which would allow Boeing to assemble these complete sections within three days at its plant located in Everett, Washington (Figure A.5).

We now explain the rationale behind the 787's supply chain as highlighted in Table A.3.

Outsource more

By outsourcing 70 per cent of the development and production activities under the 787 programme, Boeing can shorten the development time by leveraging suppliers' ability of developing different parts in parallel. Also, Boeing may be able to reduce the development cost of

Table A.3 Comparison of Boeing's strategy for 737 and 787 programmes

Component	737 programme	787 programme
Sourcing strategy	Outsourced 35–50%	Outsourced 70%
Supplier relationship	Traditional supplier relationship (purely contract based)	Strategic partners with tier-1 suppliers
Supplier responsibilities	Developed and produced parts for Boeing	Developed and produced sections for Boeing
Number of suppliers	Thousands	Approximately 50 tier-1 strategic partners
Supply contracts	Fixed price contracts with delay penalty	Risk sharing contracts
Assembly operations	30 days for Boeing to perform final assembly	Three-day assembly of complete sections

the 787 by exploiting suppliers' expertise. As Boeing outsourced more, communication and coordination between Boeing and its suppliers became critical for managing the progress of the 787 development programme. To facilitate the coordination and collaboration among suppliers and Boeing, Boeing implemented a web-based tool, Exostar, that is intended to gain supply chain visibility, improve control and integration of critical business processes, and reduce development time and cost (Manufacturing Business Technology, 2007).

Reduce direct supply base, delegate more and focus more

To reduce development time and cost for the Dreamliner, Boeing fostered strategic partnerships with approximately 50 tier-1 suppliers who will design and build entire sections of the plane and ship them to Boeing. By reducing its direct supply base, Boeing could focus more of its attention and resources on working with tier-1 suppliers (pre-integration stages) rather than with raw material procurement and early component subassembly. (The benefits of reducing the direct supply base have been used to explain how Toyota managed to develop new cars faster and cheaper than GM in the automotive industry. The reader is referred to Dyer and Ouchi (1993) for details. However, unless the supplier relationship is managed correctly, reducing the supply base can increase supply risks because of the reduced bargaining power of the manufacturer (Tang, 1999.) The rationale behind this shift is to empower its strategic suppliers to develop and produce different sections in parallel so as to reduce the development time. Also, by shifting more assembly operations to its strategic partners located in different countries, there is a potential saving in development cost.

Reduce financial risks

Under the 787 programme, Boeing instituted a new 'risk sharing' contract under which no strategic suppliers will receive the payment for the development cost until Boeing delivers its first 787 to its customers (slated to be ANA airlines). This contract payment term was intended to provide incentives for strategic partners to collaborate and coordinate their development effort. While this contract imposes certain financial risks for Boeing's strategic suppliers if delivery deadlines are missed, they are incentivized by being allowed to own their intellectual property, which can then be licensed to other companies in the future. Another incentive for the strategic partners to accept this payment term is because it allows them to increase their revenues (and potential profits) by taking up the development and production of the entire section of the plane instead of a small part of the plane.

Increase production capacity without incurring additional costs

Decentralizing the manufacturing process would allow Boeing to outsource non-critical processes. The intention is to reduce the capital investment for the 787 development programme. Also, under the 787 supply chain, Boeing needs only three days to assemble complete sections of the Dreamliner at its plant. Relative to the 737 supply chain, this drastic reduction in cycle time would in turn increase Boeing's production capacity without incurring additional investments.

New product development and design risks

While the 787 supply chain (Figure A.4) has great potential for reducing development time and cost, there are various underlying supply chain risks. As described in Sodhi and Tang (2009a), there are many types of supply chain risks, ranging from technology to process risks, from demand to supply risks and from IT system to labour risks. In this section, we shall present some of the risks and actual events that caused major delays in the Dreamliner's development programme (Table A.4).

The 787 Dreamliner involves the use of various unproven technologies. Boeing encountered the following technical problems that led to a series of delays:

Table A.4 Boeing's 787 supply chain risks and consequences

Risk factor	Potential risk caused by the 787 supply chain	Risk consequence: what happened at Boeing?
Technology	Infeasibility of material in flight tests, which is untested on this scale	Invisibility of development issues with tier-1 suppliers' partners resulting in major delays
Supply	Tier-1 suppliers outsource development tasks to tier-2 partners, which may not have technical know-how	Lack of knowledge about supplier selection by tier-1 partner, delay in development and manufacturing work
Process	Over-reliance on tier-1 partners to coordinate their development tasks with their suppliers further down the supply chain	Need for increased coordination of suppliers' activities required 'travelled work' by Boeing personnel
Management	Inexperienced management team without supply chain expertise	Management failure, need for reorganization at highest levels
Labour	Union dissatisfaction with Boeing's decision to outsource more	Union strike causing work stoppage
Demand (customer)	Publicity of problems may cause problems with airline and passenger perceptions of Boeing	Delivery delays may cause financial penalties and cancellation of orders

- **Composite fuselage safety issues:** the Dreamliner contains 50 per cent composite material (carbon fibre reinforced plastic), 15 per cent aluminium and 12 per cent titanium. The composite material has never been used on this scale and many fear that creating an aircraft with this mixture of materials is not feasible. Also, lightning strikes are a safety concern for wings made out of this composite material because a lightning bolt would potentially travel through wing-skin fasteners (Wallace, 2006b).

- **Engine interchangeability issues:** one of the key benefits of the 787's modular design concept was to allow airlines to use two different types of engines (Rolls-Royce and GE) interchangeably. Due to recent technical difficulties and part incongruity, it would take 15 days to change engines from one model to the other instead of the intended 24 hours (Leeham Co, 2005).

- **Computer network security issues:** the current configuration of electronics on the Dreamliner puts passenger electronic entertainment on the same computer network as the flight control system. This raises a security concern for terrorist attacks (Zetter, 2008).

Supply risks

Boeing is relying on its tier-1 global strategic partners to develop and build entire sections of the Dreamliner that are based on unproven technology. Any break in the supply chain can cause significant delays of the overall production. In early September 2007, Boeing announced a delay in the planned first flight of the Dreamliner citing ongoing challenges including parts shortages and remaining software and systems integration activities. Even with a web-based planning system, Exostar, used to coordinate the supplier development activities, coordination is only possible when accurate and timely information is provided by different suppliers. For example, one of the tier-1 suppliers, Vought, hired Advanced Integration Technology (AIT) as a tier-2 supplier to serve as a system integrator without informing Boeing. AIT is supposed to coordinate with other tier-2 and tier-3 suppliers for Vought (Tang, 2007). Additionally, due to cultural differences, some tier-2 or tier-3 suppliers do not often enter accurate and timely information into the Exostar system. Consequently, various tier-1 suppliers and Boeing were not made aware of the technical problems (such as parts shortages) or delay problems in a timely fashion.

Process risks

The underlying design of the 787 supply chain is likely to cause major delays because its efficiency depends on the synchronized just-in-time deliveries of all major sections from Boeing's tier-1 strategic partners. If the delivery of a section is delayed, the delivery schedule of the aircraft is delayed. Unless Boeing keeps some safety stocks of different complete sections, it is likely for Boeing to face late delivery. Also, under the risk-sharing contract, none of the strategic partners will get paid until the first completed plane is certified for flight. As strategic partners recognize the potential of being penalized unfairly if they complete their tasks before other suppliers, the risk-sharing contract payment may actually entice these strategic partners to work slower, which undermines the original intent of the risk-sharing contract (Kwon *et al*, 2009).

Management risks

As Boeing used an unconventional supply chain structure to develop and build its Dreamliner, it is essential for Boeing to assemble a leadership team that includes some members who have a proven supply chain management record with expertise to prevent and anticipate certain risks as well as to develop contingency plans to mitigate the impact of different types of risks. However, Boeing's original leadership team for the 787 programme did not include members with expertise on supply chain risk management. Without the requisite skills to manage an unconventional supply chain, Boeing was undertaking a huge risk in uncharted waters.

Labour risks

As Boeing increased its outsourcing effort, Boeing workers became concerned about their job security. Their concerns resulted in a strike by more than 25,000 Boeing employees starting in September of 2008. The effects of the worker strike were also felt by Boeing's strategic partners. For example, anticipating that the strike at Boeing would trigger order cancellations and delivery delay of certain Boeing aircraft, Spirit Aerosystems, a key supplier of Boeing, reduced its work week for employees who develop and manufacture various Boeing aircrafts. This reduced work schedule could potentially delay the delivery schedule of certain fuselage parts for the 787 (Rigby and Hepher, 2008).

Demand risks

As Boeing announced a series of delays, some customers lost their confidence in Boeing's aircraft development capability. In addition, there is a growing concern about the fact that the first 787s are over-weight by about 8 per cent, or 2.2 metric tons, which can lead to a 15 per cent reduction in range (Norris, 2009). In response to Boeing's production and delivery delays and the doubt about 787's long-range capability, some customers have begun cancelling orders for the Dreamliner or migrating toward leasing contracts instead of purchas-ing the aircraft outright. As of July 2009, the orders for the Dreamliner have been reduced from 895 (reported in November 2008) to 850 (reported in July 2009). The reader is referred to Sanders (2009a) for details.

Boeing's reactive risk mitigation strategy

To manage various disruptions as presented in the previous section, we now present Boeing's reactive response for reducing the negative impact of the current problems and for avoiding further complica-tions resulting in additional delays (Table A.5).

Mitigating technology risks

To improve the safety of its composite fuselage, Boeing is redesign-ing its fuselage by using additional material to strengthen the wing structure; however, this additional material would increase the

Table A.5   Boeing's reactive risk mitigation strategies

Risk factor	Reactive risk mitigation strategy
Technology	Modify design
Supply	Purchase company at the bottleneck stage (Vought Aircraft Industries)
Process	Boeing sends hundreds of its engineers to solve issues with underperforming partners at their sites
Management	Reorganization of top management – replaced programme manager with supply chain expert
Labour	Concessions to labour unions – increased pay and decreased outsourcing
Demand (customer)	Boeing to pay penalties for delivery delays, public relations campaign to reassure customers

aircraft's overall weight. Boeing management has continued to assure its customers that it would work diligently to reduce the weight of the final version of the plane. Regarding the time it takes to change engines from one model to the other, Boeing is redesigning its installation process with the hope of reducing its changeover time. Finally, to ensure that the computer network is secure, a proper design is required that allows for the separation of the navigation computer systems from the passenger electronic entertainment system.

Mitigating supply risks

After realizing that some tier-1 strategic partners did not have the know-how to develop different sections of the aircraft nor experience in managing their tier-2 suppliers to develop the requisite components for the sections, Boeing recognized the need to regain control of the development process of 787. For instance, knowing that Vought Aircraft Industries was the weakest link in the Boeing's 787 supply chain, Boeing acquired one unit of Vought in 2008 and then another unit in 2009 (Ray, 2008; Sanders, 2009b). These two acquisitions would provide Boeing direct control of these two units of Vought and their tier-2 suppliers for the fuselage development. Further, as a result of continued production delays, some of Boeing's suppliers were in jeopardy of facing massive profit losses, which put completion of the entire Dreamliner programme at risk. For example, in response to threats of work stoppage, Boeing paid its tier-1 strategic partner, Spirit Aerosystems, approximately US $125 million in 2008 to ensure that this partner continued its vital operations (Rigby, 2008).

Mitigating process risks

As a response to suppliers' inability to meet production deadlines, Boeing decided that it must send key personnel to sites across the globe to fill suppliers' management vacuum and address production issues in person. This proved to be an expensive endeavour as personnel were pulled from responsibilities on-site at Boeing to address supply and manufacturing issues at the sites of their outsourced partners. The strategy of relying on suppliers for subassembly proved to be too risky for Boeing in certain circumstances and resulted in Boeing having to perform the work themselves. For instance, Boeing sent hundreds of its engineers to the sites of various tier-1, tier-2, or tier-3 suppliers worldwide to solve various technical problems that appeared to be the root cause of the delay in the 787 development. Ultimately, Boeing had to redesign the entire aircraft subassembly process (Gunsalus, 2007). While this 'hands on' approach would certainly help, it is very costly and time consuming, which defeats the underlying intents of the 787's redesigned supply chain as described earlier.

Mitigating management risks

To restore customers' confidence about Boeing's aircraft development capability and to reduce any further delays, Boeing recognized the need to bring in someone with a proven record of supply chain management expertise. In response, the original 787 programme director, Mike Bair (with proven marketing expertise), was replaced by Patrick Shanahan who had proven expertise in supply chain management. In his new role, Shanahan is now responsible for coordination of all activities for Boeing's major plane families, which includes the Dreamliner. Moreover, Boeing has changed it top leadership by replacing its Interim CEO, James Bell, with Jim McNerney in 2005.

Mitigating labour risks

To bring about an end to the strike after two months of shutdown, Boeing made concessions that would give workers a 15 per cent wage boost over four years. On the key issue of job security, which had been the major impediment to reaching an agreement, Boeing agreed to limit the amount of work that outside vendors could perform. Therefore, Boeing's concept of outsourcing a significant amount of work to global partners could be endangered and production costs could eventually rise. In response to the wage increases and limits in outsourcing promised by Boeing, the machinists union conceded to withdraw charges filed with the Department of Labor regarding allegations of unfair bargaining practices at Boeing (Gates, 2008).

Mitigating demand (customer) risks

As customers have begun to cancel their 787 orders and as the company's capability of developing the 787 is in question, Boeing has developed the following mitigation strategies. First, as a way to compensate its customers' potential loss due to the late deliveries of their orders, Boeing is supplying replacement aircraft (new 737 or 747) to various concerned airlines such as Virgin Atlantic (Lunsford, 2007; Crown, 2008). Second, to restore Boeing's public image, Boeing has improved its communication by sharing its progress updates on its webpage. In addition, Boeing is conducting a publicity campaign to promote the superior technology of the planes and the overall value that the aircraft will offer to airlines and passengers (Crown, 2008).

Boeing's potential proactive risk mitigation strategies

As Boeing makes its best effort to restore confidence in its capability of developing innovative aircrafts such as the Dreamliner, there are certain risk mitigation strategies that Boeing could have embarked on at the outset of the programme to better manage potential risks proactively (Table A.6).

Improve supply chain visibility

As described earlier, Boeing's supply risk was caused by the lack of supply chain visibility. Without accurate and timely information about the supply chain structure and the development progress at

Table A.6 Alternative strategies for mitigating programme risks

Risk factor	Proactive actions	Risk effect
Supply chain visibility	Use IT to ensure transparency of entire supply chain	Avoided or reduced
Strategic partner selection and relationship	Proper vetting of all strategic partners to determine their capability of completing tasks	Reduced
Process	Develop better risk sharing opportunities and incentives for strategic partners	Reduced
Management	Establish proper working team with expertise in supply chain logistics	Avoided
Labour	Outreach and communication with union heads to discuss sourcing strategies	Avoided
Demand (customer)	Treat customers as partners and better communicate the potential for missing delivery deadlines	Avoided or reduced

each supplier's site, the value of Exostar has been compromised significantly. To improve information accuracy, Boeing should have required that all strategic partners and suppliers provide all information embedded in the supply chain relationships instead of relying on alerts generated from the programme only when they were directly affected. Also, Boeing should provide incentives for all suppliers to use Exostar to communicate accurate information in a timely manner.

Improve strategic supplier selection process and relationships

Spending more effort on evaluating each supplier's technical capability and supply chain management expertise for developing and manufacturing a particular section of the Dreamliner would have enabled Boeing to select more capable tier-1 strategic suppliers, which could avoid or reduce potential delays caused by inexperienced tier-1 suppliers. Also, Boeing should require participation in the tier-1 partner's vetting process of tier-2 (or tier-3) suppliers. The additional effort of properly vetting key suppliers would certainly enhance communication and coordination and reduce the risks of potential delays, which would in turn reduce the development time and cost (Lunsford, 2007).

Modify the risk-sharing contract

While the delayed payment term associated with the risk-sharing contract was intended to reduce Boeing's financial risk, it did not provide proper incentives for tier-1 suppliers to complete their tasks early. When some strategic partners are incapable of developing their sections according to the plan schedule, the entire development schedule is pushed back. As a result of these delays, Boeing incurred millions of dollars in penalties that it had to pay out to its customers (West, 2007). To properly align the incentives among all strategic partners, Boeing should have structured the contracts with reward (penalty) for on-time (late) delivery (Kwon et al, 2009).

Proactive management team

Boeing should have chosen the right people for the job at the outset of the programme, allowing them to avoid and anticipate the risks associated with its novel supply chain structure. Also, identifying the sources of potential problems and having the right person (or team) in place would mitigate many of the risks, and allow Boeing to respond more quickly and effectively when problems occurred. For example, Boeing could have either avoided or anticipated various types of supply chain risks as described earlier had they appointed persons with proven supply chain management expertise to serve on the original leadership team. By having a leadership team with all requisite skills, Boeing would have the requisite expertise and authority to respond to the delay problems more effectively.

Proactive labour relationship management

Dissatisfaction among Boeing's machinists was caused by Boeing's strategy to increase its outsourced operations to external suppliers. Had the union's general disapproval of Boeing's outsourcing strategy been taken into account, Boeing may not have decided to outsource 70 per cent of its tasks. Even if this outsourcing strategy was justified financially, Boeing could have managed its labour relationship proactively by discussing the strategy, by offering job assurances and by obtaining buy-in from unions. This proactive labour relationship management would have created a more mutually beneficial partnership, which could have avoided the labour strikes.

Proactive customer relationship management

Recognizing the risks associated with innovative product development, proactive customer relationship management is critical to help customers set proper expectations when placing their orders. Better communication with customers throughout the development process can enable a company to manage customers' perceptions throughout the entire product development process. Setting proper expectations about the delivery schedules of its 787 Dreamliner may have encouraged the airlines to manage their aircraft replacement schedule differently, say order more 737s and 747s and fewer 787s. Without setting an aggressive delivery schedule to its customers, it was plausible for Boeing to reduce the penalty caused by the delayed delivery schedule. Through continuous engagement and open communication about the challenges and Boeing's contingency plans, it would have been plausible for Boeing to manage its customers' perception and its reputation better.

Afterthoughts

Through our description of its complex supply chain earlier in the chapter, our analysis of various supply chain risks and Boeing's reactive mitigation strategies discussed in the previous section, we have developed the following insights that other manufacturers may consider when managing their supply chains for efficient new product development.

Assembling a leadership team with requisite expertise

On the surface, it appears that Boeing's fundamental problem was caused by its attempts to simultaneously take on too many drastic changes. These changes include unproven technology, unconventional supply chains, unproven suppliers' capability to take on new roles and responsibilities, and unproven IT coordination systems. However, one plausible reason for Boeing to take on so many drastic changes may be because the 787 leadership team underestimated the risks associated with all these changes. Had Boeing constructed a multi-disciplinary team with expertise to identify and evaluate various supply chain risks, it would have been quite possible for Boeing to avoid and anticipate potential risks, and to develop proactive mitigation strategies and contingency plans to reduce the impact of various supply chain disruptions.

Obtaining internal support proactively

Partnerships between management and labour are essential for smooth operations for companies to implement any new initiatives including new product development programmes. While their interests are often misaligned, better communication of business strategies with union workers is likely to be a proactive step towards avoiding costly worker strikes. Also, aligning the incentives for both parties proactively is more likely to reduce potential internal disruptions down the road.

Improving supply chain visibility to facilitate coordination and collaboration

Besides the need to perform due diligence in key supplier selection to ensure that the selected supplier has the requisite capability and the commitment for success, a company should consider cultivating stronger commitment in exchange for accurate information in a timely manner. Overly relying on IT communication is highly risky when managing a new project. To mitigate the risks caused by partners further upstream or downstream, companies should strive to gain complete visibility of the entire supply chain. Having clear supply chain visibility would enhance the capability for a company to take corrective action more quickly, which is more likely to reduce the negative impact of a disruption along the supply chain. The reader is referred to Sodhi and Tang (2009b) for a discussion of the importance of timely response to mitigate the negative effects of supply chain disruptions.

Proactive management of customer expectation and perception

Due to the inherent risks associated with new product development, it is critical for a company to help its customers set proper expectations proactively, especially regarding the potential delay caused by various types of risks as highlighted in the tables above. Setting proper expectations at the outset would reduce potential customer dissatisfaction down the road. During the development phase, it is advisable for the company to maintain open and honest communication with its customers regarding the actual progress, technical challenges and corrective measures. Such efforts would possibly gain customer trust, which would improve their loyalty in the long run.

Conclusion

Boeing's Dreamliner programme involved dramatic shifts in supply chain strategy from traditional methods used in the aerospace industry. In addition, Boeing boasted about its novel manufacturing techniques and its technological marvels. Such dramatic shifts from convention involve significant potential for encountering risks throughout the process. Boeing's ongoing issues with meeting delivery deadlines are a direct result of its decision to make drastic changes in the design, the development process and the supply chain associated with the Dreamliner programme simultaneously without having the proper management team in place. Further, this team did not proactively assess the risks that were later realized and did not develop coherent strategies for effectively mitigating them. While it may be impossible to identify all potential risks and create contingency plans for all eventualities before a project begins, Boeing could have done many things differently. It is instructive for managers in any industry to view the issues that Boeing faced and analyse how these issues were handled so that they can learn from mistakes that were made before engaging in similar supply chain restructuring.

Epilogue

When Boeing announced its plan to develop the Dreamliner in 2003, it was intended to deliver its first plane within four years. However, due to a series of mishaps ranging from shortage of fasteners, incomplete software, technical problems associated with the horizontal stabilizers and electrical fires during test flights, etc, the delivery of the first Dreamliner to ANA took place in September 2011 (which was more than three years behind schedule). Also, it was reported that the actual development and manufacturing cost was US $32 billion (which exceeds the original budget by US $26 billion). In addition to over-budget and way behind schedule, the Dreamliner has a series of technical problems. In 2012 and early 2013, white smoke and fire were reported by ANA and JAL, and fuel leaks were reported by ANA and JAL in 2013. Due to safety concerns, all 50 Dreamliners were grounded by the FAA until Boeing developed a solution in late 2013. Since late 2013, no major issues have been reported. The reader is referred to Reuters (2013) for details. In hindsight, Boeing could have reduced or even avoided these problems had they managed their design (internally) and their supply chain design more carefully.

Notes

1 UCLA Anderson School, 110 Westwood Plaza, Los Angeles, CA 90095, USA; Email: ctang@anderson.ucla.edu. This chapter is based on the materials presented in Tang and Zimmerman (2009).

2 Other immediate customers include air freight logistics service providers such as Federal Express or DHL and aircraft operators such as Global Air.

References

Boeing website, www.boeing.com

Crown, J (2008) Will Boeing Pay for Delays? Spiegel Online International, April 4 [Online] available at: http://www.spiegel.de/international/business/0,1518,545365,00.html [accessed May 2009]

Dyer, J and Ouchi, W (1993) Japanese Style Business Partnerships: Giving companies a competitive edge, *Sloan Management Review*, 35 (1), pp 51–63

Gates, D (2008) Boeing strike ends; Machinists back on the job Sunday, *Seattle Times*, 2 November [Online] available at: http://seattletimes. nwsource.com/html/localnews/2008340022_webmachinists02m.html [accessed April 2009]

Gunsalus, J (2007) Boeing Sticks to Revised 787 Dreamliner Schedule, *Bloomberg Press*, 11 December [Online] available at: http://www. bloomberg.com/apps/news?pid=20601103&sid=aqgUCtUslurM&refer =us [accessed April 2009]

Hawk, J (2005) The Boeing 787-Dreamliner: More than an airplane, May [Online] available at: http://www.aiaa.org/events/aners/Presentations/ ANERS-Hawk.pdf [accessed April 2009]

Hucko, C (2007) Airbus A380 vs. Boeing 787: Poll reveals that passengers prefer a smaller plane, Suite101, 28 April [Online] available at: http://airplanes.suite101.com/article.cfm/airbus_a380_vs_boeing_787 [accessed May 2009]

Kwon, D, Lippman, SA, McCardle, K and Tang, CS (2009) Time Based Contracts with Delayed Payments, working paper, UCLA Anderson School

Leeham Co (2005) 787 Is not Meeting 24hour-Engine Change Promo, lessor says, Leeham Co, LLC, 18 July [Online] available at: http://www. leeham.net/filelib/SCOTTSCOLUMN071805.pdf [accessed May 2009]

Lunsford, J (2007) Boeing, in Embarrassing Setback, Says 787 Dreamliner will be Delayed, Wall Street Journal, 11 October

Manufacturing Business Technology (2007) Boeing 787 program flying smoothly with Exostar collaboration engine, *Manufacturing Business Technology*, 1 March [Online] available at: http://www.mbtmag.com/ article/CA6420965.html [accessed April 2009]

Murray, C (2007) Boeing 787 Dreamliner Rolls Out Smoother Ride with Gust Suppression, Design News, 4 June [Online] available at: http://www.designnews.com/article/439-Boeing_787_Dreamliner_Rolls_ Out_Smoother_Ride_with_Gust_Suppression.php [accessed April 2009]

Norris, G (2009) 787s Move Along, Weight Problems Persist, *Aviation Daily*, 6 May [Online] available at: http://www.aviationweek.com/ aw/generic/story_generic.jsp?channel=aviationdaily&id=news/787 NEW05059.xml [accessed May 2009]

Raman, A, Schmidt, C and Gaul, V (2008) Airbus A380 – Turbulence Ahead, Harvard Business School Case, Number N9-609-041

Ray, S (2008) Boeing Buys Vought Venture to Stem Delays on 787, *Bloomberg Press*, 28 March [Online] available at: http://www.bloomberg.com/apps/news?pid=20601103&sid=aw4dIEC3nhGs&refer=news [accessed April 2009]

Reuters (2013) Timeline: Boeing's 787 Dreamliner woes, 12 July [Online] available at: http://www.reuters.com/article/us-boeing-dreamliner-timeline-idUSBRE96B13L20130712 [accessed August 2016]

Rigby, B (2008) Spirit Aero to get Early Payments from Boeing, Reuters,10 April [Online] available at: http://uk.reuters.com/article/basicIndustries/idUKN0946545920080410 [accessed April 2009]

Rigby, B and Hepher, T (2008) Boeing Strike Impact to be Felt Globally, Reuters, September [Online] available at: http://www.reuters.com/article/ousivMolt/idUSN0529451820080908 [accessed 22 July 2009]

Sanders, P (2009a) Boeing Tightens Its Grip on Dreamliner Production, *Wall Street Journal*, B1, 2 July

Sanders, P (2009b) Boeing Sets Deal to Buy a Dreamliner Plant, *Wall Street Journal*, 8 July

Sanders, P (2009c) Boeing Yet to Clear Dreamliner's Takeoff, *Wall Street Journal*, B3, 23 July

Sodhi, M and Tang CS (2009a) Managing supply chain disruption via time-based risk management, in *Managing Supply Chain Risk and Vulnerability: Tools and Methods for Supply Chain Decision Makers*, eds Blackhurst and Wu, Springer

Sodhi, M and Tang, CS (2009b) Supply Chain Risk Management, in *Encyclopedia of Operations Research and Management Science*, eds Cochran, Anthony, Keskinocak, Kharoufeh and Smith, John Wiley & Sons

Tang, CS (1999) Supplier Relationship Map, International Journal of Logistics Research and Applications, 2 (1), pp 39–56

Tang, CS (2007) Boeing's 787 Supply Chain: A dream or a nightmare?, unpublished paper, UCLA Anderson School

Tang, CS and Zimmerman, JD (2009) Managing New Product Development and Supply Chain Risks: The Boeing 787 case, *Supply Chain Forum: An International Journal*, 10 (2), pp 74–86

Wallace, J (2006a) Airbus Unveils Widebody, says A350 XWB will top 787 and 777, *SeattlePI*, 18 July [Online] available at: http://www.seattlepi.com/business/277877_airshow18.html [accessed April 2009]

Wallace, J (2006b) Aerospace Notebook: Lightning a Weighty Issue for the 787, *SeattlePI*, 12 July [Online] available at: http://www.seattlepi.com/business/277220_air12.html [accessed April 2009]

West, K (2007) Boeing 787 Program not out of Woods, MSNBC, 12 December

Zetter, K (2008) *FAA: Boeing's New 787 May be Vulnerable to Hacker Attack, Wired,* 8 January [Online] available at: http://www.wired.com/politics/security/news/2008/01/dreamliner_security [accessed May 2009]

GLOSSARY

agility One of three key attributes of any supply chain, which also includes flexibility and strength. The most mature definition of agility is the ability to recalculate plans in the face of the market, demand and supply volatility and deliver the same or comparable cost, quality and customer service.

brand loyalty The tendency of some consumers to continue buying the same brand of goods, rather than competing brands.

business process re-engineering (BPR) The analysis and redesign of workflows within and between enterprises in order to optimize end-to-end processes and automate non-value-added tasks.

change management The management of change and development within a business.

concurrent engineering Also known as simultaneous engineering, a method of designing and developing products, in which the different stages run simultaneously, rather than consecutively.

constrained design The practice of limiting user actions on a system. Constraints limit the actions that can be performed by the user, thus increasing the usability of the design and reducing the likelihood of operator error.

continuity of supply A cohesive and ongoing system that distributes a product.

customer configuration Software that manages information regarding customer accounts.

downstream supply chain Downstream refers to where products are produced and distributed.

inventory holdings The storing of remaining products.

long-distance offshore manufacture Relocation of an operational process from one country to another.

manufacturability Also known as design for manufacturing or DFM, the general engineering practice of designing products in such a way that they are easy to create.

remanufacturing   Where end-of-life products are returned to the original manufacturer, stripped down or disassembled, with individual components being – where possible – refurbished and then reassembled into a complete product.

sales revenues   Income from sales of goods and services, minus the cost associated with things like returned or undeliverable merchandise.

space utilization   A measure of whether and how space is being used. The utilization rate is a function of a frequency rate and an occupancy rate.

supply chain strategy   An iterative process that evaluates the cost–benefit trade-offs of operational components.

supply chain vulnerability   A point of weakness or possible threat to the supply chain network.

sustainability   An issue affecting an organization's supply chain or logistics network in terms of environmental, risk and waste costs.

transport intensity   Concerns the economic or energy efficiency of transport.

upstream supply chain   Upstream refers to the material inputs needed for production.

REFERENCES

2degrees (nd) ASDA Case Study, 2degrees network [Online] available at: https://www.2degreesnetwork.com/2degrees/downloads/asda-case-study. pdf [accessed 29 January 2018]

APICS (2018) SCOR – The Most Recognized Supply Chain Framework [Online] available at: http://www.apics.org/apics-for-business/products-and-services/apics-scc-frameworks/scor [accessed 31 January 2018]

Australian Aluminium Council (2018) *FAQs* [Online] available at: https://aluminium.org.au/faqs/ [accessed 1 February 2018]

BBC News (1999) Business: The Company File – Sparks at Marks [Online] available at: http://news.bbc.co.uk/1/hi/business/285252.stm [accessed 12 January 2018]

BBC News (2017) Tesla's market value overtakes Ford [Online] available at: http://www.bbc.co.uk/news/business-39485200 [accessed 11 January 2018]

Berdichevsky, G, Kelty, K, Straubel, JB and Toomre, E (2006) The Tesla Roadster Battery System [Online] available at: http://large.stanford.edu/publications/power/references/docs/tesla.pdf [accessed 11 January 2018]

Bessant, J, Neely, AD, Tether, B, Whyte, J and Yaghi, B (2006) *Intelligent Design: How managing the design process effectively can boost corporate performance*, Advanced Institute of Management Research, London

Bidault, F, Despres, C and Butler, C (1998) New Product Development and Early Supplier Involvement (ESI), *International Journal of Technology Management*, **15** (1/2), pp 46–69

Boeing website, www.boeing.com

Brass, C, Bowden, F and Moseley, J (2007) *Design for Sustainable Change: a positioning paper*, Design Council, London

Burke, WW and Litwin, GH (1992) A Causal Model of Organizational Performance and Change, *Journal of Management*, **18** (3), 1 September, pp 523–45

Burkett, Michael J (2006) Design for Supply: An evolutionary path: design for supply involves more than just part and supplier reuse.

Leading companies see benefits from simulation and modelling before and during product launch (Technology), *Supply Chain Management Review*, Peerless Media [retrieved 15 January 2018 from HighBeam Research: https://www.highbeam.com/doc/1G1-155615409.html]

Cameron, E and Green, M (2015) *Making Sense of Change Management: A complete guide to the models, tools and techniques of organisational change*, 4th edn, Kogan Page, London

Cargille, B and Fry, C (2006) Design for Supply Chain: Spreading the word across HP [Online] available at: https://kelley.iu.edu/mabert/e730/HP-Design_SCMR_2006.pdf [accessed 15 January 2018]

Carson, R (1962) *Silent Spring*, Houghton Mifflin Company

Christopher, M (1998) *Logistics and Supply Chain Management: Strategies for reducing cost and improving service*, 2nd edn, FT-Prentice Hall, London, p 49

Christopher, M (2005) *Logistics and Supply Chain Management: Creating value adding networks*, 3rd edn, Prentice Hall, London

Christopher, M (2011) *Logistics and Supply Chain Management: Creating value-adding networks*, 4th edn, Financial Times Prentice Hall, London

Clark, KB and Fujimoto, T (1991) *Product Development Performance: Strategy, organization, and management in the world auto industry*, Harvard Business School Press, Boston

Clark, N (2006) The Airbus Saga: Crossed wires and a multibillion-euro delay – Business – International Herald Tribune, *The New York Times* [Online] available at: http://www.nytimes.com/2006/12/11/business/worldbusiness/11iht-airbus.3860198.html [accessed 12 January 2018]

Cooper, R (2001) *Winning at New Products*, 3rd edn, Perseus Books, Cambridge, MA

Cooper, RG and Edgett, SJ (2006) Stage-Gate® and the Critical Success Factors for New Product Development, *BPTrends*, July [Online] available at: http://www.bptrends.com/publicationfiles/07-06-ART-Stage-Gate-ForProductDev-Cooper-Edgett1.pdf [accessed 19 January 2018]

Cox, G (2005) *Cox Review of Creativity in Business: Building on the UK's strengths*, HMSO [Online] available at: http://webarchive.nationalarchives.gov.uk/20120704143146/http://www.hm-treasury.gov.uk/d/Cox_review-foreword-definition-terms-exec-summary.pdf [accessed 31 January 2018]

Crown, J (2008) Will Boeing Pay for Delays? Spiegel Online International, April 4 [Online] available at: http://www.spiegel.de/international/business/0,1518,545365,00.html [accessed May 2009]

Design Council (2004) *Design in Britain*, Internal report by the Design Council

Design Council (2007) *The value of design factfinder report* [online] available at: https://www.designcouncil.org.uk/sites/default/files/asset/document/TheValueOfDesignFactfinder_Design_Council.pdf [accessed 15 May 2018]

Design Council (2011) *Design for Innovation* [Online] available at: https://www.designcouncil.org.uk/sites/default/files/asset/document/design-for-innovation.pdf [accessed 11 January 2018]

Design Council, Warwick Business School (2014) *Leading Business by Design: Why and how business leaders invest in design* [Online] available at: https://www.designcouncil.org.uk/sites/default/files/asset/document/dc_lbbd_report_08.11.13_FA_LORES.pdf [accessed 11 January 2018]

Dittman JP (2014) *Managing Risk in the Global Supply Chain: A report by the supply chain management faculty at the University of Tennessee*, The Global Supply Chain Institute, The University of Tennessee College of Business Administration

Dyer, J and Ouchi, W (1993) Japanese Style Business Partnerships: Giving companies a competitive edge, *Sloan Management Review*, 35 (1), pp 51–63

Ellen MacArthur Foundation (2015) *Towards a Circular Economy: The business rationale for an accelerated transition* [Online] available at: https://www.ellenmacarthurfoundation.org/assets/downloads/TCE_Ellen-MacArthur-Foundation_9-Dec-2015.pdf [accessed 30 January 2018]

Eveleigh, C (2009) If you cannot measure it, you cannot improve it, *Getting to Excellent* [Online] available at: http://www.gettingtoexcellent.com/2009/05/if-you-cannot-measure-it-you-cannot.html [accessed 31 January 2018]

Ferdows, K, Machuca, J and Lewis, M (2015) *Zara: The World's Largest Fashion Retailer*, case study [Online] available at: https://www.thecasecentre.org/corporate/products/view?id=130606 [accessed 15 January 2018]

Fine, CH, Golany, B and Naserald, H (2005) Modelling Tradeoffs in Three-Dimensional Concurrent Engineering: A goal programming approach, *Journal of Operations Management,* **23** (3–4), pp 389–403

Forbes (2016) *Forbes* Global 2000 [Online] available at: https://www.forbes.com/global2000/#7e1deab2335d [accessed 11 January 2018]

Fuller, T (2011) Thailand Flooding Cripples Hard-Drive Suppliers, *The New York Times* [Online] available at: http://www.nytimes.com/2011/11/07/business/global/07iht-floods07.html [accessed 19 January 2018]

Ganapathy, B and Goh, CH (1997) A Hierarchical System of Performance Measures for Concurrent Engineering, *Concurrent Engineering: Research and Application* **5** (2), pp 137–43

Gates, D (2008) Boeing Strike Ends: Machinists back on the job Sunday, *Seattle Times,* 2 November [Online] available at: http://seattletimes.nwsource.com/html/localnews/2008340022_webmachinists02m.html [accessed April 2009]

GlaxoSmithKline (2015) Press Release, 2degreesnetwork [Online] available at: https://www.2degreesnetwork.com/2degrees/downloads/GSK_press release_-_14_Sept_2015.pdf [accessed 29 January 2018]

Glende, WL (1997) The Boeing 777: A look back, NATO STO, Seattle [Online] available at: https://pdfs.semanticscholar.org/0b62/33d8f075be5cb74af32968b219863ec3d703.pdf [accessed 22 January 2018]

Goffin, K and Mitchell, R (2005) The Risks of Innovation, *Financial Times,* September 30

Gunsalus, J (2007) Boeing Sticks to Revised 787 Dreamliner Schedule, *Bloomberg Press,* 11 December [Online] available at: http://www.bloomberg.com/apps/news?pid=20601103&sid=aqgUCtUslurM&refer=us [accessed April 2009]

Hammer, M (1990) Reengineering Work: Don't Automate, Obliterate, *Harvard Business Review,* July

Hansen, S (2012) How Zara Grew Into the World's Largest Fashion Retailer, *The New York Times* [Online] available at: http://www.nytimes.com/2012/11/11/magazine/how-zara-grew-into-the-worlds-largest-fashion-retailer.html [accessed 12 January 2018]

Hawk, J (2005) The Boeing 787-Dreamliner: More than an Airplane, May [Online] available at: http://www.aiaa.org/events/aners/Presentations/ANERS-Hawk.pdf [accessed April 2009]

Henley Forum (nd) *Henley Forum - Knowledge in action leaflets* [Online] available at: https://www.henley.ac.uk/research/research-centres/the-henley-forum-for-organisational-learning-and-knowledge-strategies/folks-knowledge-in-action [accessed 31 January 2018]

Henley Forum (nd) Knowledge in Action Issue 32 [Online] available at: https://www.henley.ac.uk/research/research-centres/the-henley-forum-for-organisational-learning-and-knowledge-strategies/folks-knowledge-in-action [accessed 12 January 2018]

Hoskins, T (2017) H&M, Zara and Marks & Spencer linked to polluting viscose factories in Asia, *The Guardian*, 13 June [Online] available at: https://www.theguardian.com/sustainable-business/2017/jun/13/hm-zara-marks-spencer-linked-polluting-viscose-factories-asia-fashion [accessed 30 January 2018]

HP, Inc (2016) *2016 Sustainability Report* [Online] available at: http://www8.hp.com/h20195/v2/GetPDF.aspx/c05507473.pdf [accessed 30 January 2018]

Hucko, C (2007) Airbus A380 vs. Boeing 787: Poll Reveals that Passengers Prefer a Smaller Plane, Suite101, 28 April [Online] available at: http://airplanes.suite101.com/article.cfm/airbus_a380_vs_boeing_787 [accessed May 2009]

Isaacson, W (2015) *Steve Jobs*, Abacus, London

Jackson-Proes, A (2001) Trouble Deepens as M&S Sales Decline, *The Telegraph* [Online] available at: http://www.telegraph.co.uk/finance/2725809/Trouble-deepens-as-Marks-and-Spencer-sales-decline.html [accessed 12 January 2018]

Johnsen, TE (2009) Supplier Involvement in New Product Development and Innovation: Taking stock and looking to the future, *Journal of Purchasing and Supply Management*, **15** (3), pp 187–97

Johnson, ME (2002a) Product Design Collaboration: Capturing lost supply chain value in the apparel industry (April), Tuck School of Business Working Paper No. 02-08 [Online] available at SSRN: https://ssrn.com/abstract=307461 or http://dx.doi.org/10.2139/ssrn.307461 [accessed 15 January 2018]

Johnson, ME (2002b) Product Design Collaboration: Capturing lost supply chain value in the apparel industry, *Textile Digest*, November

Kaye, L (2013) Clothing to Dye For: The textile sector must confront water risks, *The Guardian*, 12 August [Online] available at: https://

www.theguardian.com/sustainable-business/dyeing-textile-sector-water-risks-adidas [accessed 30 January 2018]

Keller, J (2005) Lead-Free Solder: A train wreck in the making, *Military Aerospace* [Online] available at: http://www.militaryaerospace.com/articles/print/volume-16/issue-10/news/trends/lead-free-solder-a-train-wreck-in-the-making.html [accessed 19 January 2018]

Khan, O and Creazza, A (2009) Managing the Product Design–Supply Chain Interface: Towards a roadmap to the design centric business, *International Journal of Physical Distribution and Logistics Management*, **39** (4), pp 301–19

Khan, O and Cutler-Greaves, Y (2008) Mitigating Supply Chain Risk through Improved Agility: Lessons from a UK retailer, *International Journal of Agile Systems and Management*, **3** (3–4), pp 263–81

Kotter, J (1995) Leading Change: Why transformation efforts fail, *Harvard Business Review*, March–April, **73** (2) pp 59–67, (republished January 2007)

Kotter, J (2012) *Leading Change*, Harvard Business Review Press, Brighton, MA

Kwon, D, Lippman, SA, McCardle, K and Tang, CS (2009) Time Based Contracts with Delayed Payments, working paper, UCLA Anderson School

Lee, H (2004) The Three A's of Supply Chain Excellence, *EE Times* [Online] available at: https://www.eetimes.com/document.asp?doc_id=1240026 [accessed 24 January 2018]

Lee, H, Billington, C and Carter, B (1993) Hewlett-Packard Gains Control of Inventory and Service through Design for Localization, *Interfaces*, **23** (4), pp 1–11

Leeham Co (2005) 787 Is not Meeting 24 hour-Engine Change Promo, lessor says, Leeham Co, LLC, 18 July [Online] available at: http://www.leeham.net/filelib/SCOTTSCOLUMN071805.pdf [accessed May 2009]

Lubben, R (1988) *Just in Time Manufacturing: An aggressive manufacturing strategy*, McGraw-Hill, New York

Lunsford, J (2007) Boeing, in Embarrassing Setback, Says 787 Dreamliner will be Delayed, *Wall Street Journal*, 11 October

Manufacturing Business Technology (2007) Boeing 787 Program Flying Smoothly with Exostar Collaboration Engine, *Manufacturing Business Technology*, 1 March [Online] available at: http://www.mbtmag.com/article/CA6420965.html [accessed April 2009]

Meadows, DH, Meadows, DL, Randers, J and Behrens III, WW (1972) *The Limits to Growth: A report for the Club of Rome's project on the predicament of mankind*, Universe Books, ISBN 0-87663-165-0

Murphy, J (2008) What a Bright Idea: Innovation stems from convergence of design, supply chain excellence, *Supply Chain Brain* [Online] available at: http://www.supplychainbrain.com/index.php?id=7098&type=98&tx_ttnews%5Btt_news%5D=3909&cHash=da03e20e36 [accessed 19 January 2018]

Murray, C (2007) Boeing 787 Dreamliner Rolls Out Smoother Ride with Gust Suppression, *Design News*, 4 June [Online] available at: http://www.designnews.com/article/439-Boeing_787_Dreamliner_Rolls_Out_Smoother_Ride_with_Gust_Suppression.php [accessed April 2009]

New Look (2018) [Online] available at: http://www.newlookgroup.com/who-we-are/our-brand [accessed 15 January 2018]

Nguyen, H, Stuchtey, M and Zils, M (2014) Remaking the Industrial Economy, *McKinsey Quarterly*, February [Online] available at: http://thebusinessleadership.academy/wp-content/uploads/2016/03/Circular_economy.pdf [accessed 30 January 2018]

Norris, G (2009) 787s Move Along, Weight Problems Persist, *Aviation Daily*, 6 May [Online] available at: http://www.aviationweek.com/aw/generic/story_generic.jsp?channel=aviationdaily&id=news/787NEW05059.xml [accessed May 2009]

Paternoga, RK (2015) Caterpillar's Remanufacturing Business Helps Make Sustainable Progress Possible, in *Achieving a Circular Economy: How the private sector is reimagining the future of business* [Online] available at: https://www.uschamberfoundation.org/sites/default/files/Circular%20Economy%20Best%20Practices.pdf [accessed 30 January 2018]

Pearce, DW and Turner, RK (1989) *Economics of Natural Resources and the Environment*, Johns Hopkins University Press, Baltimore, MD

Pich, MT, Van Der Heyden, L and Harle, L (2002) *Marks & Spencer and Zara: Process competition in the textile apparel industry* [Online] available at: https://hbr.org/product/marks-spencer-and-zara-process-competition-in-the-textile-apparel-industry/INS849-PDF-ENG [accessed 12 January 2018]

Poirier, C and Walker, I (2005) *Business Process Management Applied: Creating the value-managed enterprise*, J Ross Publishing, Boca Raton, FL

Raman, A, Schmidt, C and Gaul, V (2008) Airbus A380 – Turbulence ahead, Harvard Business School Case, Number N9-609-041

Ray, S (2008) Boeing Buys Vought Venture to Stem Delays on 787, *Bloomberg Press*, 28 March [Online] available at: http://www.bloomberg.com/apps/news?pid=20601103&sid=aw4dIEC3nhGs&refer=news [accessed April 2009]

Reuters (2013) Timeline: Boeing's 787 Dreamliner woes, 12 July [Online] available at: http://www.reuters.com/article/us-boeing-dreamliner-timeline-idUSBRE96B13L20130712 [accessed August 2016]

Rigby, B (2008) Spirit Aero to get Early Payments from Boeing, *Reuters*, 10 April [Online] available at: http://uk.reuters.com/article/basicIndustries/idUKN0946545920080410 [accessed April 2009]

Rigby, B and Hepher, T (2008) Boeing Strike Impact to be Felt Globally, *Reuters*, September [Online] available at: http://www.reuters.com/article/ousivMolt/idUSN0529451820080908 [accessed 22 July 2009]

Rudzki, R and Trent, R (2011) *Next Level Supply Management Excellence*, J Ross Publishing, Fort Lauderdale, FL

Sanders, P (2009a) Boeing Tightens Its Grip on Dreamliner Production, *Wall Street Journal*, B1, 2 July

Sanders, P (2009b) Boeing Sets Deal to Buy a Dreamliner Plant, *Wall Street Journal*, 8 July

Sanders, P (2009c) Boeing Yet to Clear Dreamliner's Takeoff, *Wall Street Journal*, B3, 23 July

Seetharaman, D (2011) *Automakers Face Paint Shortage after Japan Quake*, Reuters [Online] available at: https://www.reuters.com/article/us-japan-pigment/automakers-face-paint-shortage-after-japan-quake-idUSTRE72P04B20110326 [accessed 1 February 2018]

Senge, P, Kleiner, A, Roberts, C, Ross, R, Roth, G and Smith, B (1999) *The Dance of Change: The challenges to sustaining momentum in a learning organization*, Nicholas Brealey Publishing, London

Sodhi, M and Tang, CS (2009a) Managing Supply Chain Disruption via Time-Based Risk Management, in *Managing Supply Chain Risk and Vulnerability: Tools and methods for supply chain decision makers*, eds Blackhurst and Wu, Springer

Sodhi, M and Tang, CS (2009b) Supply chain risk management, in *Encyclopedia of Operations Research and Management Science*, eds Cochran, Anthony, Keskinocak, Kharoufeh and Smith, John Wiley & Sons

Tang, CS (1999) Supplier Relationship Map, *International Journal of Logistics Research and Applications*, **2** (1), pp 39–56

Tang, CS (2007) Boeing's 787 Supply Chain: A dream or a nightmare?, unpublished paper, UCLA Anderson School

Tang, CS and Zimmerman, JD (2009) Managing New Product Development and Supply Chain Risks: The Boeing 787 case, *Supply Chain Forum: An International Journal*, **10** (2), pp 74–86

Tencer, D (2017) BlackBerry's Market Share Declines To 0.0%, Huffington Post Canada [Online] available at: http://www.huffingtonpost. ca/2017/02/16/blackberry-market-share_n_14798544.html [accessed 11 January 2018]

The Coca-Cola Company (2017) *Water Stewardship: 2016 Sustainability Report* [Online] available at: http://www.coca-colacompany.com/ stories/2016-water-stewardship [accessed 30 January 2018]

The Economist (2001) Marks's Failed Revolution [Online] available at: http://www.economist.com/node/569731 [accessed 11 January 2018]

The Economist (2017) The Bad Earth: The most neglected threat to public health in China is toxic soil, 8 June, pp 24–6 [Online] available at: https://www.economist.com/news/briefing/21723128-and-fixing-it-will-be-hard-and-costly-most-neglected-threat-public-health-china [accessed 30 January 2018]

The International Aluminium Institute (2017) *Aluminium in Packaging – Lightweight* [Online] available at: http://packaging.world-aluminium. org/benefits/lightweight/ [accessed 30 January 2018]

US Chamber of Commerce Foundation (2015) *Achieving a Circular Economy: How the Private Sector Is Reimagining the Future of Business* [Online] available at: https://www.uschamberfoundation. org/sites/default/files/Circular%20Economy%20Best%20Practices.pdf [accessed 30 January 2018]

Wallace, J (2006a) Airbus Unveils Widebody, says A350 XWB will top 787 and 777, *SeattlePI*, 18 July [Online] available at: http://www.seattlepi. com/business/277877_airshow18.html [accessed April 2009]

Wallace, J (2006b) Aerospace Notebook: Lightning a Weighty Issue for the 787, *SeattlePI*, 12 July [Online] available at: http://www.seattlepi.com/ business/277220_air12.html [accessed April 2009]

Walton, M (2015) Inside Sony Design: How a new breed of designers hope to fix the company's fortunes, Ars Technica [Online] available at:

http://arstechnica.co.uk/gadgets/2015/10/inside-sony-design-how-a-new-breed-of-designers-hope-to-fix-the-companys-fortunes/ [accessed 11 January 2018]

West, K (2007) Boeing 787 Program Not Out of Woods, *MSNBC*, 12 December

Wheatley, M (1998) Pile It Low, Sell It Fast, *Management Today*, 1 February, p 62

Wheatley, M (1999) Compaq vs Dell, *Logistics Europe*, September, p 18

Wheatley, M (2005) Boosting Productivity, *The Manufacturer* [Online] available at: https://www.themanufacturer.com/author/malcolm-wheatley/ [accessed 15 January 2018]

Wheatley, M (2008) Simulation, Better Forecasting, and Advanced Planning Smooth Lean Transition to the Supply Chain, *Manufacturing Business Technology*, January

Wheatley, M (2009) Simulating the Factory, *Engineering and Technology*, August

Wheatley, M (2010) Awash with Ideas, *Procurement Leaders*, (24), pp 28–31

Wheatley, M (2011) After the Disaster in Japan, *Automotive Logistics* [Online] available at: http://automotivelogistics.media/intelligence/after the-disaster [accessed 19 January 2018]

Wheatley, M (2013) Virtual Reality, *The Manufacturer*, 4 September, p 70

Whyte, J, Bessant, J and Neely, AD (2005) Management of Creativity and Design within the Firm, paper commissioned by the Department of Trade and Industry as an input to the Creativity Review [Online] available at: http://webarchive.nationalarchives.gov.uk/20111108232833/http://www.bis.gov.uk/files/file14795.pdf [accessed 11 January 2018]

Winner, RI, Pennell, JP, Bertrand, HE and Slusarczuk, MM (1988) *The Role of Concurrent Engineering in Weapons System Acquisition*, Institute for Defense Analyses, Alexandria, VA

Womack, J, Jones, D and Roos, D (1990) *The Machine that Changed the World*, Simon & Schuster, London

Zetter, K (2008) *FAA: Boeing's New 787 May Be Vulnerable to Hacker Attack, Wired*, 8 January [Online] available at: http://www.wired.com/politics/security/news/2008/01/dreamliner_security [accessed May 2009]

FURTHER READING

Baer, D (2014) The Making of Tesla: Invention, Betrayal, and the Birth of the Roadster, Business Insider [Online] available at: http://uk.businessinsider.com/tesla-the-origin-story-2014-10 [accessed 11 January 2018]

Fine, C, Padurean, L and Werner, M (2014) The Tesla Roadster (A): Accelerated supply chain learning, Massachusetts Institute of Technology Sloan School of Management [Online] available at: http://asb.edu.my/wp-content/themes/asb/Tesla%20Roadster%20(A)%20 2014.pdf [accessed 11 January 2018]

Johansson, H and Carr, D (1995) Best Practices in Reengineering: What works and what doesn't in the reengineering process, McGraw-Hill, London

Matthews, P and Syed, N (2004) The Power of Postponement, Supply Chain Management Review, 8 (3), pp 28–34

Serling, RJ (1992) Legend and Legacy: The story of Boeing and its people, St. Martin's Press, London

Siemens PLM Software (2016) Product Lifecycle Management Software [Online] available at: https://www.plm.automation.siemens.com/en_us/plm/ [accessed 15 January 2018]

Sigaria (2014) Knowledge share, A Procurement Leaders 'Insight' paper

Spitz, W, Golaszewski, R, Beradino, F and Johnson, J (2001) Development Cycle Time Simulation for Civil Aircraft, NASA/CR-2001-210658, pp 2–7

INDEX

Note: page numbers in *italic* indicate figures or tables.

KOGAN PAGE
LOGISTICS

*Cutting-edge insights
from leading thinkers in
academia and industry*

Organizations now consider packaging to be a critical issue. Effective and efficient packaging can significantly improve the performance of companies and minimize costs. *Packaging Logistics* examines all the essential roles within an organization, from the purchasing of raw materials to the production and sale of finished products, as well as transport and distribution. It is important for practitioners and academics to understand the role of packaging along the supply chain and the different implications of an efficient product packaging system for the successful management of operations.

The book covers hot topics such as sustainability, innovation, returns, e-commerce, end-of-life and future trends and challenges. With more than 30 case studies, *Packaging Logistics* takes the reader through every stage of packaging and relates it to the supply chain and logistics.

HENRIK PÅLSSON

PACKAGING LOGISTICS

UNDERSTANDING AND MANAGING THE ECONOMIC AND ENVIRONMENTAL IMPACTS OF PACKAGING IN SUPPLY CHAINS

KoganPage

Henrik Pålsson is Associate Professor at Lund University, Sweden. He has conducted research and taught Master's students in packaging logistics since 2004.

PACKAGING LOGISTICS

www.koganpage.com/
packaginglogistics

Leading the way in current thinking on environmental logistics, *Green Logistics* provides a unique insight into the environmental impacts of logistics and the actions that companies and governments can take to deal with them. It is written by leading researchers in the field and provides a comprehensive view of the subject for students, managers and policymakers.

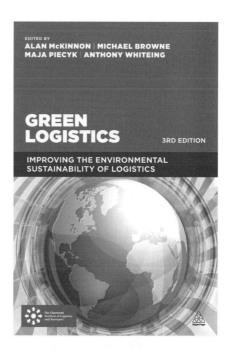

Alan McKinnon is Professor of Logistics at Kühne Logistics University, Hamburg. He has been researching and teaching freight transport and logistics for almost forty years and has published extensively in journals and books.

Michael Browne is Professor of Logistics at the University of Westminster, London. He has worked on studies for Transport for London, the European Commission, the UK Department for Transport, research councils and commercial organizations.

Maja Piecyk is Associate Professor in Logistics at Heriot-Watt University, Edinburgh. Much of her research focuses on the optimization of supply networks, carbon auditing of logistics and long-term trends in the energy requirements and environmental impacts of logistics.

Anthony Whiteing is Senior Lecturer in the Institute for Transport Studies at the University of Leeds. An academic with 30 years' experience, he's been involved in various UK and European research projects primarily in the field of freight transport.

GREEN LOGISTICS

www.koganpage.com/GL

Decarbonizing Logistics outlines the nature and extent of the challenge we face in trying to achieve deep reductions in greenhouse gas emissions from logistical activities. It makes a detailed assessment of the available options, including restructuring supply chains, shifting freight to lower-carbon transport modes and transforming energy use in the logistics sector. The options are examined from technological and managerial standpoints for all the main freight transport modes.

Based on an up-to-date review of almost 600 publications and containing new analytical frameworks and research results, *Decarbonizing Logistics* is the first book to provide a global, multidisciplinary perspective on the subject. It is written by one of the foremost specialists in the field.

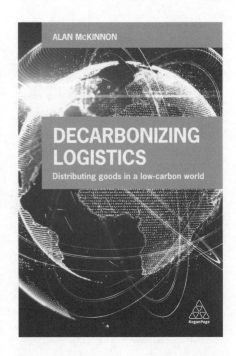

ALAN McKINNON

DECARBONIZING LOGISTICS

Distributing goods in a low-carbon world

KoganPage

Alan McKinnon is Professor of Logistics at Kühne Logistics University, Hamburg. He has been researching and teaching freight transport and logistics for almost forty years and has published extensively in journals and books. He was a member of the European Commission's High Level Group on Logistics, Chairman of the World Economic Forum's Logistics and Supply Chain Industry Council and a lead author of the transport chapter of the Intergovernmental Panel on Climate Change's Fifth Assessment Report. He has spent many years researching the links between logistics and climate change and been an adviser to governments, international organizations and companies.

DECARBONIZING LOGISTICS

www.koganpage.com/DL

MAGNUS CARLSSON

STRATEGIC SOURCING AND CATEGORY MANAGEMENT

LESSONS LEARNED AT IKEA

Magnus Carlsson has twenty-five years' experience as a strategic sourcing expert at IKEA. He developed and led the implementation of IKEA's purchasing strategy, which included category-based sourcing. He was responsible for the strategic sourcing of IKEA's global textile, metal and veneer businesses, as well as component and raw material procurement. He is currently a lecturer, trainer and adviser in strategic sourcing and category management.

Strategic Sourcing and Category Management examines how category management works in practice, drawing insight from IKEA. With over twenty-five years' experience of purchasing at IKEA, Magnus Carlsson shares the wisdom gained from devising and executing IKEA's highly successful purchasing strategies. The text is supported by insightful interviews and case studies, which are compared and contrasted with examples from other leading companies.

Strategic Sourcing and Category Management answers three critical questions: When is category management a profitable method and why? How do category teams create real results? How can category management be organized and implemented effectively?

In answering these three questions, Magnus Carlsson not only presents the guiding principles behind category-based sourcing, he also demonstrates how category-based sourcing can effectively be implemented in practice and provides guidance on how to realize the benefits of this approach. With application spanning far beyond IKEA, this book is an invaluable resource for procurement professionals in any industry.

STRATEGIC SOURCING AND CATEGORY MANAGEMENT

www.koganpage.com/SSCM

Over the last two decades Tesco has emerged as a dominant player in the UK market and a leading global retailer. The second edition of *The Lean Supply Chain* explores how Tesco, over the last 20 years or so, has built its business around supply chain excellence. As a megaretailer, Tesco has learnt to create a balanced supply chain system, supporting suppliers' needs as well as customers' requirements. This perspective, and an ambition to act sustainably, has underpinned a rebuilding of trust in the Tesco brand and a resurgence in commercial fortunes.

The first edition of *The Lean Supply Chain* was highly acclaimed, winning the prestigious *Prix des Associations* 2016, in *Les Plumes des Achats & Supply Chain*. In this new edition, which contains new chapters on Tesco's current strategy, rebuilding trust in the brand and the CSR agenda, the authors chart the principles of Lean thinking, customer loyalty and simplicity which were used by Tesco to frame its supply chain strategy. They draw upon their deep knowledge of how the retailer has dealt with challenges and market changes from both academic and practitioner perspectives to provide lessons for other businesses, large or small, who wish to place how they manage their supply chains at the heart of their competitive strategy.

Barry Evans has held roles in Tesco including Lean Process Manager in Tesco Supply Chain Development. He joined the Lean Enterprise Research Centre at Cardiff Business School as a Senior Research Associate.

Robert Mason is a Reader in the Logistics and Operations Management section at Cardiff Business School and has led numerous business research projects, including with Tesco as a partner.

THE LEAN
SUPPLY CHAIN

www.koganpage.com/LSC

A Circular Economy Handbook for Business and Supply Chains is an easily digestible and comprehensive handbook that provides a clear guide to the circular economy, helping the reader create future-fit, sustainable strategies. Real examples across a range of market sectors help businesses, students and policymakers understand the theory and fast-developing practice of the circular economy. To help the reader generate ideas, *A Circular Economy Handbook for Business and Supply Chains* provides a holistic framework for the design and supply chain and supporting business models, and includes tools the reader can use to get started.

While growing global consumption presents fantastic business opportunities, our current linear systems (take some materials, make a product, use it and then throw it away) are not fit for purpose. The circular economy unlocks this problem by decoupling resources from consumption. Engaged businesses are rethinking product design, material choices, business models and supply chains. This book is a must-read for anyone who wants to apply the circular economy today.

Catherine Weetman helps businesses develop future-proofed, resilient strategies, assessing sustainability risks and value opportunities. She is a Visiting Fellow at the University of Huddersfield, a Vice-Chair of the Environment and Sustainability Forum at the Chartered Institute of Logistics and Transport and gained an MSc in Logistics from Cranfield University. Her background includes industrial engineering in manufacturing and retail distribution, logistics solution design, project management, business intelligence, logistics product development and supply chain consulting.

A CIRCULAR ECONOMY HANDBOOK FOR BUSINESS AND SUPPLY CHAINS

www.koganpage.com/CEH

Business Operations Models provides a strategic framework through which business leaders can understand and identify the potential in their operations for achieving super-performance and possible disruptor status. It tries to bridge the gap between business strategy thinking and operations management. Research presented in the book points to the characteristics and drivers of disruptive competitors: remarkable perseverance, serendipity, burning platforms and overcoming disbelief. Using detailed international case studies, *Business Operations Models* explores in detail how winning competitors work, including:

- the power of 1% in eliminating waste

- leveraging technology for market transformation

- redefining channels to create disruptive business models

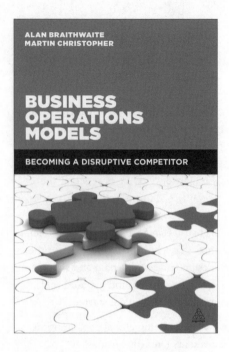

Alan Braithwaite is chairman and founder of LCP Consulting, which has worked with over 400 companies internationally. He is also a visiting professor at Cranfield University and specializes in supply chain strategy and operational excellence in the retail, manufacturing and service sectors.

Martin Christopher has been at the forefront of the development of logistics and supply chain management for over 30 years. He is Emeritus Professor of Marketing and Logistics at Cranfield University, where he helped build the Centre for Logistics and Supply Chain Management.

BUSINESS OPERATIONS MODELS

www.koganpage.com/ busopmod